Applied Reliability for Industry 3

Reliability of Multiphysical Systems Set

coordinated by
Abdelkhalak El Hami

Volume 18

Applied Reliability for Industry 3

Operational Reliability for the Automobile, Aeronautics, Defense, Medical, Marine and Space Industries

Edited by

Abdelkhalak El Hami
David Delaux
Henri Grzeskowiak

iSTE

WILEY

First published 2023 in Great Britain and the United States by ISTE Ltd and John Wiley & Sons, Inc.

ISTE Ltd
27-37 St George's Road
London SW19 4EU
UK

www.iste.co.uk

John Wiley & Sons, Inc.
111 River Street
Hoboken, NJ 07030
USA

www.wiley.com

Library of Congress Control Number: 2022946026

British Library Cataloguing-in-Publication Data
A CIP record for this book is available from the British Library
ISBN 978-1-78630-693-7

Contents

· their remarkable work in the implementation of six volumes in all the authors who have devoted many hours to shaping all their ; meant that this essential information does not remain the property ed by the largest number of engineers, technicians, researchers and

lume is devoted to Predictive Reliability, for estimating or predicting erformance of systems.

l volume is dedicated to Experimental Reliability using tools and test monstrate the reliability of systems.

volume presents Operational Reliability; it verifies the reliability of systems in real life by way of an analysis of the field data.

ike to thank all those involved in this program as well as the financiers ons) without whom these projects could not have succeeded and I wish nbers of the hard core set up in these programs (AUDACE, FIRST-MFP, IOS, SYMSTECK) continue their association because their skills will be r guaranteeing the reliability of the very many technologies which are in s of emerging and which will equip the mechatronics systems of

Philippe EUDELINE
President of Normandie AeroEspace (NAE)
October 2022

Predicting and then guaranteeing the reliabi challenge for manufacturers in the automotive, a well as for those in the rail, telecommunicat among many others. But above all, it is a maj equipment on a day to day basis, who must information transmitted and the decisions mac development of connected objects (autonomous c lead to a drastic reduction of human interventior mechatronic systems. All these systems can only be confidence in the reliability of the equipment.

These forms of equipment are divided into two n part mainly composed of electronic boards (coupled w real time software part enabling the implementatio realization of the expected tasks.

Predicting and guaranteeing the reliability of electro endless task. On the one hand, the number of compone boards is very high and on the other hand, the new fi equipment require numerous reliability and robustness tes

Thanks to the support of the Astech and MOV'EO co NAE aeronautics industry, the Normandy and Ile de Franc and Versailles Chambers of Commerce, programs have bee produced exceptional results (AUDACE, FIRST-M SYMSTECK, etc.).

Given the richness of the results of these three books that would like to thank very warmly Abdelkhalak El Hami, Da

Grzeskowiak fo total as well as results. This ha of few but shar students.

The first vo the reliability

The secon methods to de

The third performance

I would (State, Regi that the me PISTIS, CR essential fo the proces tomorrow.

Preface

March 2020 was a turning point in our lives, for industry worldwide, etc., when Covid-19 struck the world! This pandemic has highlighted the *fragility* of our industries while exacerbating the paradigms of development, innovation, commercial competition and the acceleration of research phases.

More than ever, we need to be *agile* in our American, European, Asian and even global business plans, while at the same time *cementing* the use of our products in their operational use. It is clear that the *challenges* of combining the concepts of variability and stability, standardization and acceleration of product development are enormous. The "post-Covid" world is pushing all industries to the very limits of reliability engineering.

It is true that the word "reliability" is often used as a marketing cosmetic without any real care for its underlying scientific substantiation! Nonetheless, behind this word lies an applied science, validated by analysis, research and talented people.

Which type of the reliability approach can a designer, an engineer and a manager use in their professional environment? Whatever the industrial field (aeronautics, space, defense, automotive, home automation, etc.), how is it possible to design a reliable product more quickly, all the while upholding safety measures?

This book illustrates, in great detail, the state of the art for the multidisciplinary science of operational reliability. The authors wish to clearly depict this discipline through concrete examples from the industry, based on user experience feedback.

1

Durability Approach: Applied to a Vehicle Lighting Control System

Ensuring the durability of system components is a strategic activity for companies in key industrial sectors, such as transportation. The stakes are twofold: economic impact and brand image. Indeed, a lack of control over the reliability of a product can lead to significant costs. This is becoming all the more of a reality given the recent trend to increase the length of the warranty period [DEL 16]. Although it is not the only factor, the reliable or durable design of components contributes to these costs, and thus their improvement leads to a reduction. However, beyond the direct economic impact, the brand image of the product is also affected. This is difficult to quantify objectively. However, a customer, whether a private individual or a professional, who encounters problems with their vehicle before its resale period (which is on average 5 years of use worldwide [CCF 19]) will naturally be disappointed and may not choose to make their next purchase from the same brand. However, in a sector as competitive as the automotive industry of today, a bad or good image can change our place in the market in a significant way.

1.1. Introduction

Designing a reliable car requires knowledge of information, such as the customer's expected usage requirements (region, country, type of use and changes in mobility solutions), as well as the strength of the parts that make up the car. It is also necessary to consider all the constraints and evolutions of the component, such as

For a color version of all the figures in this chapter, see http://www.iste.co.uk/elhami/applied3.zip.

Chapter written by Medoune NDIAYE and Caroline RAMUS-SERMENT.

the mass (which has an important impact on consumption) or the choice of materials (associated with the recyclability of products).

In this chapter, a global approach to sustainability will be illustrated through the example of a vehicle lighting control system. This approach is similar to the one presented in the guide "*Aide à la conception et validation de la fiabilité automobile*"[1] [SIA 19]. It is based on the concepts of reliability and probabilities:

- quality requirements at mid-life and/or over lifetime;
- technical standards and methodological tools (stress–strength method);
- reliability management plan between manufacturers and suppliers.

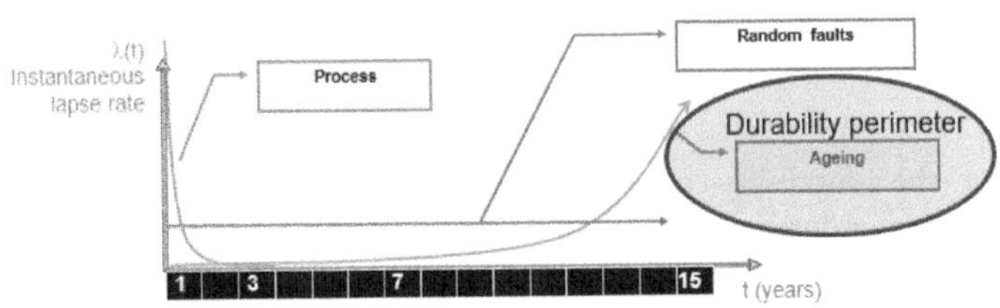

Figure 1.1. *Instantaneous rate of failure over time and study of the durability perimeter*

As shown in Figure 1.1, a component is subject to different causes of failure which occur in the early-stage of lifetime (less than one year of usage). These failures represent pathologies to which the component does not resist for a long time before showing symptoms. By definition, they are caused by insufficient burn-in or a faulty process.

- Random failures: Under these failures, several of the actual causes of failure are hidden. These are generally due to a lack of robustness in terms of the design or the process.

- Wear-and-tear failures: These failures are characteristic for the end of a component's life. They are the result of reaching the physical and chemical limits of the materials used.

1.2. Example of a vehicle lighting control system

The lighting control of a vehicle allows for the activation of the lights: light position, high beam or low beam. When a driver selects the desired position, a

1 Design assistance and validation of automotive reliability.

ı the example being studied, we will focus mainly on the increase in ıistance of the contact between the moving track and the PCB.

			Comments
ight ction	Headlight loss: Increased contact resistance	*Oxidation*	*Gold does not oxidize, but copper alloy does. As long as the gold layer is present, it protects against oxidation.*
		Fretting – corrosion	*The presence of the gold layer prevents the phenomenon of frictional corrosion. If this gold layer tends to disappear, the vibrations transmitted to the component by the chassis/road will lead to corrosion of the part.*
		Mechanical breakage	*Inadequate component design can compound problems of aging.*
		Mechanical fatigue	*Track: due to bending movement of the PCB.* *Moving contact: due to repetitive bending stresses.*
		Differential thermal expansion	*Track circuit: due to temperature cycling during operation and/or ambient temperature variation (negligible).* *Track – moving contact: the contact time is not long enough to be important.*
		Thermal ageing (diffusion)	**Copper diffusion in gold tinning [PIN 79]. Once the entire thickness of the gold film is penetrated, copper accumulates on the surface and a layer of high copper concentration exists immediately below the gold–air interface.**
		Pollution	*Not relevant because IP5x component is protected against dust.*
		Wear-and-tear	**The surfaces of the moving contact and the track are rough.** **The thin gold layer is subject to frictional wear-and-tear.**

Table 1.1. *Potential versus probable failure mechanisms*

mechanical contact is established between a mov (see Figure 1.2). This mechanical contact allows t lamps (headlights).

1.2.1. *Risks and reliability requirements*

When contact is lost between the moving par (PCB) track, power is interrupted and the lights go ou safety issue.

Indeed, the untimely shutdown of the lights is ASIL B, according to several Original Equipment Man the rating rules of the ISO 26262 Standard [ISO 11].

For any event classified as ASIL B, the ISO 262(maximum probability such that this event is equivalent to

Assuming a constant in-service failure rate, this i maximum probability of 7.5E-4 of the lights going out un the failure mechanism) over a vehicle service life of approx

In addition, a clause in ISO 26262 requires that th hardware (raw materials, hardware, as opposed to soft calculations based on user experience or testing, must have least 70% [ISO 11]. For this example, we set a confidence lev

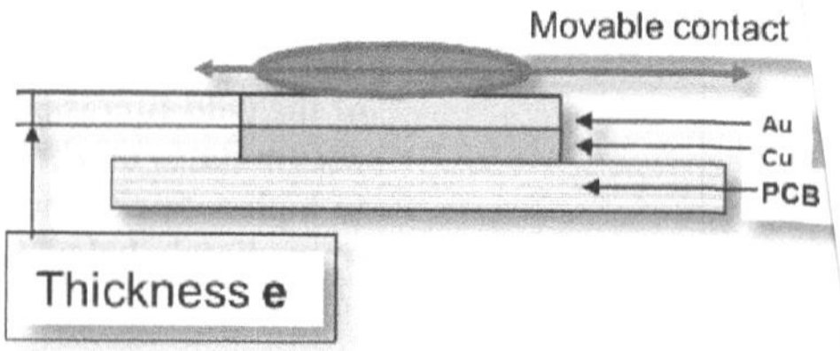

Figure 1.2. *Representation of the electrical contact made via*

1.2.2. *From failure modes to failure mechanisms*

We will now focus on the technical points. If we carry out a risk a Failure Mode Effects and Analysis (FMEA) or a Fault Tree Analys event "unexpected loss of lights", we can see, according to the tech of the light control, that several failure mechanisms [SIA 17] can lea

First, we can identify all the *potential* failure mechanisms that can occur, next we will focus on the *most likely* ones.

The most likely major failure mechanisms are listed in bold in Table 1.1.

On our PCB, we have a copper alloy track covered by a gold layer (see Figure 1.2). The gold layer prevents two main types of mechanism failures:

– oxidation of the copper alloy layer;

– corrosion of the copper alloy through friction.

However, it introduces two other failure mechanisms:

– the diffusion of copper through gold;

– the wear-and-tear of the gold layer.

In Table 1.1, the following is presented:

– in italics: the potential failure mechanisms (by no means an exhaustive list).

Here, we have two main competitive failure mechanisms:

– competitor A: diffusion of copper through the gold layer, due to thermal aging;

– competitor B: disappearance of the gold layer, due to contact wear-and-tear.

The first one that occurs defines the life of the component.

1.2.3. *From failure mechanisms to physical damage factors*

Now we need to look for the physical factors that will activate these failure mechanisms.

Failure mechanisms	Physical damage factors
Thermal aging (diffusion, Kirkendall effect, etc.)	Track temperature
Wear-and-tear	Contact force (N) and relative displacement (mm) between the PCB track and the moving contact

Table 1.2. *Failure mechanisms and physical damage factors*

wear-and-tear is often represented by a Weibull distribution. With failure test procedures, it is possible to identify the appropriate statistical distribution in order to have the values of the shape (β) and scale parameters (η).

In practice, failure tests often involve a limited number of components (<15). The fitting of the strength distribution must therefore be done with robust methods for small samples such as the median rank method [ABE 06].

The fitting procedure that can be easily performed with a spreadsheet is as follows:

– Step 1: Order the number of activations to failure found in the test in ascending order.

– Step 2: Estimate the values of the distribution function of the Weibull distribution (see illustration in Table 1.5) by:

$$\hat{F}_i = \frac{i-0.3}{N+0.3} \text{ with an accuracy to the order of } 5^{E}\text{-}2 \qquad [1.1]$$

Equation [1.1] is the median rank by the Benard's approximation or by:

$$\hat{F}_i = Beta^{-1}(0.5, i, N - i + 1) \text{ for an exact value.} \qquad [1.2]$$

Equation [1.1] is the median rank by the inverse beta law.

Here:

– i: the order of the failure;

– N: the size of the tested sample;

– $Beta^{-1}$: the inverse of the incomplete beta distribution.

In Table 1.5:

– the i values are in cells F17 to F22;

– the sample size N is in cell F15;

– the values of the number of activations to fail here come from a Weibull distribution with a shape parameter beta of 5.7 and a scale parameter eta of 68,000.

N			
6			
i	**Number of activations**	**F_i (Benard Approximation)**	**F_i Exact Beta Law**
1	51082	=(F17-0.3)/(F15+0.4)	=BETA.INVERSE(0.5;F17;F15-F17+1)
2	57828	=(F18-0.3)/(F15+0.4)	=BETA.INVERSE(0.5;F18;F15-F18+1)
3	61777	=(F19-0.3)/(F15+0.4)	=BETA.INVERSE(0.5;F19;F15-F19+1)
4	67850	=(F20-0.3)/(F15+0.4)	=BETA.INVERSE(0.5;F20;F15-F20+1)
5	75502	=(F21-0.3)/(F15+0.4)	=BETA.INVERSE(0.5;F21;F15-F21+1)
6	83229	=(F22-0.3)/(F15+0.4)	=BETA.INVERSE(0.5;F22;F15-F22+1)

Table 1.5. *Formulas for calculating median ranks in a spreadsheet*

This gives the results presented in Table 1.6.

N			
6			
i	**Number of Activations**	**F_i (Benard Approximation)**	**F_i Exact Beta Law**
1	51082	10.94%	10.91%
2	57828	26.56%	26.44%
3	61777	42.19%	42.14%
4	67850	57.81%	57.86%
5	75502	73.44%	73.56%
6	83229	89.06%	89.09%

Table 1.6. *Results of the calculation of median ranks with Benard's formula and the incomplete Beta distribution*

– Step 3: Identify the parameters of the linear regression $y_i = \ln\left[\ln\left(\frac{1}{1-F_i}\right)\right]$ and $x_i = \ln(ni)$ with ni as the numbers of activations at failure.

In Table 1.7, the F_i values calculated previously go from cell I17 to cell I22 and the numbers of activations to failures are filled in from cell G17 to cell G22.

Least squares regression on rows	
yi = ln [ln (1/(1 – Fi))]	xi = ln (ni)
=LN(LN(1/(1 – I17)))	=LN(G17)
=LN(LN(1/(1 – I18)))	=LN(G18)
=LN(LN(1/(1 – I19)))	=LN(G19)
=LN(LN(1/(1 – I20)))	=LN(G20)
=LN(LN(1/(1 – I21)))	=LN(G21)
=LN(LN(1/(1 – I22)))	=LN(G22)

Table 1.7. *Formulas for computing yi and xi for ordinary least squares linear regression*

This gives the following results in Table 1.8.

Least squares regression on rows	
yi = ln [ln (1/(1 – F_i))]	xi = ln (ni)
–2.158	10.841
–1.180	10.965
–0.603	11.031
–0.146	11.125
0.285	11.232
0.795	11.329

Table 1.8. *Results of yi and xi computation for ordinary least squares linear regression*

In accordance with the linear form of the Weibull distribution by double logarithmic transformation [ABE 06] and by identification:

– the shape parameter of the Weibull distribution corresponds to the slope of the regression line;

– the scale parameter of the Weibull distribution corresponds to $\eta = e^{\left(-\frac{originally\ ordered}{\text{slope}}\right)}$.

Table 1.9 shows the application in a spreadsheet.

Beta	=SLOPE(J17:J22;K17:K22)
Eta	=EXP(-ORIGINAL.ORDER(J17:J22;K17:K22)/K24)

Table 1.9. *Formulas for calculating Beta and Eta by the median rank method in a spreadsheet*

The range J17:J22 contains the values of *yi*, and the range K17:K22 contains the values of *xi*.

The results obtained are presented in Table 1.10.

Parameter	Estimated value	True value
Beta	5.8	5.8
Eta	71,156	68,000

Table 1.10. *Results estimation of beta and eta by the median ranks method in a spreadsheet*

Remember that it is extremely important to design the failure test correctly in order to avoid the activation of failure mechanisms other than the one we are trying to characterize. For example, during the wear-and-tear test, high-frequency activations should be avoided as this would lead to abnormal heating of the contact.

1.2.6. *Resistance distribution chart*

In the absence of failure tests, which can be costly in the design phase, it is however possible to define the properties that the strength distribution must respect in order for the component to be compatible with the reliability requirements and the mission profile.

This is done by constructing a chart (see Figure 1.4) that defines a level line, that is, an acceptance boundary corresponding to the maximum allowable failure probability and defined by unique pairs of the shape parameter β and the scale parameter η.

Therefore, for each value of β (image of the strength dispersion), the minimum value of the scale parameter η (image of the average strength) ensuring a failure probability equal to the target is calculated.

This is done by solving the following equation for fixed $P_{reliability}$, μ and σ fixed:

$$P_{reliability} = \int_{xmin}^{xmax} f_{stress}(x;\mu;\sigma).\,F_{strength}(x;\beta;\eta)\ \cdot dx \qquad [1.3]$$

Equation [1.3] is the probability of failure calculation by strength–stress method.

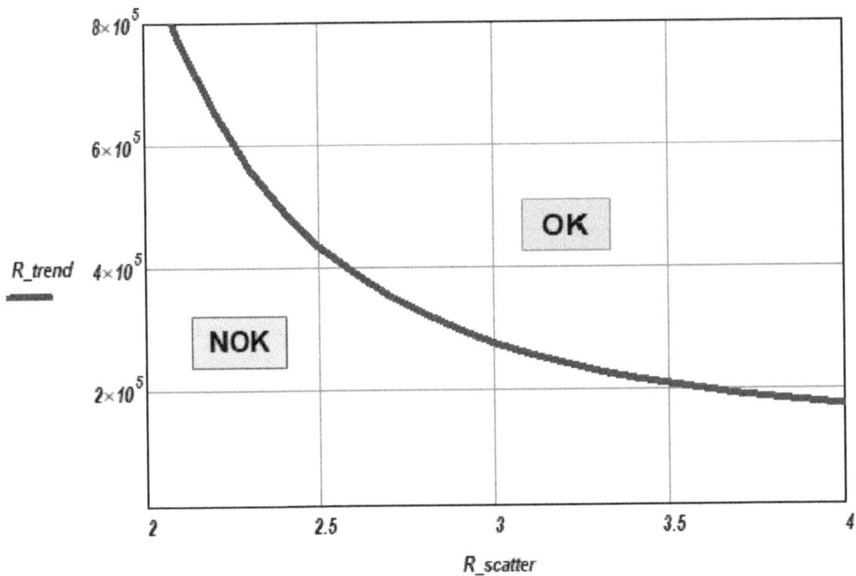

Figure 1.4. *Resistance distribution chart*

Here,

$$f_{stress}(x;\mu;\sigma) = \frac{1}{\sqrt{2\pi}\sigma x} e^{-\left(\frac{\ln(x)-\mu}{\sigma}\right)^2}$$

$$F_{strength}(x;\beta;\eta) = 1 - e^{-\left(\frac{x}{\eta}\right)^{\beta}}$$

This calculation of the integral of equation [1.3] can be easily implemented in a spreadsheet using a quadrature method of the Simpson, trapezoid, rectangle or midpoint type, as well as the statistical functions LAW.LOGNORMAL and LAW.WEIBULL.

We then simply use the spreadsheet solver to find the value of η which, for a given value of β, satisfies the equality of equation [1.3].

This results in the construction of an acceptance/rejection boundary that allows both the customer and the supplier to estimate the impact of the design (average gold layer thickness) and the process (gold layer thickness dispersion) for the achievement of reliability objectives.

Indeed, the relative strength dispersion, which is the ratio between the strength dispersion and the average strength, is expressed as:

$$q = \frac{standard\ deviation}{average} \qquad [1.4]$$

For a Weibull law, we get:

$$q = \frac{\sqrt{\eta^2 \cdot \Gamma\left(1+\frac{2}{\beta}\right) - \left(\eta \cdot \Gamma\left(1+\frac{1}{\beta}\right)\right)^2}}{\eta \cdot \Gamma\left(1+\frac{1}{\beta}\right)} = \frac{\sqrt{\Gamma\left(1+\frac{2}{\beta}\right) - \left[\Gamma\left(1+\frac{1}{\beta}\right)\right]^2}}{\Gamma\left(1+\frac{1}{\beta}\right)} \qquad [1.5]$$

Equation [1.5] is the relative dispersion as a function of beta, with the Gamma function Γ.

The application of the formula from equation [1.5] in a spreadsheet (with the value of β in cell A1) is given as:

Q = ROOT((GAMMA(1 + (2/A1)) – GAMMA(1 + 1/A1)^2))/GAMMA(1 + 1/A1)

The relationship between q and β can be correctly approximated by:

$$\begin{cases} for\ \beta \in [1 - 10[\ : q = 0.989375 \cdot \beta^{-0.9119754} \\ for\ \beta \in [10 - 100[\ : q = 1.167036 \cdot \beta^{-0.9801508} \end{cases}$$

	1	2
1	2	$9.318 \cdot 10^5$
2	2.1	$7.726 \cdot 10^5$
3	2.2	$6.534 \cdot 10^5$
4	2.3	$5.623 \cdot 10^5$
5	2.4	$4.913 \cdot 10^5$
6	2.5	$4.35 \cdot 10^5$
7	2.6	$3.897 \cdot 10^5$
8	2.7	$3.527 \cdot 10^5$
9	2.8	$3.223 \cdot 10^5$
10	2.9	$2.969 \cdot 10^5$
11	3	$2.755 \cdot 10^5$
12	3.1	$2.574 \cdot 10^5$
13	3.2	$2.419 \cdot 10^5$
14	3.3	$2.285 \cdot 10^5$
15	3.4	$2.169 \cdot 10^5$
16	3.5	...

Table 1.12. *Column 1 = beta value; Column 2 = eta value*

Weibull's law: cases where the failure mechanism was the wear-and-tear of the gold layer.

$$\textit{Distribution function of the Weibull law}, F(z) = 1 - e^{\left(-\left(\frac{z}{\eta}\right)^{\beta}\right)}$$

$$N = \frac{\ln(1-c)}{1-F(z)-1}$$

Excel ®: Nb_test = LN (1 – 0.75)/LN (1 – WEIBULL.LAW.N (115,550; 3.3; 228,500; TRUE)) – 1 = 12.2

Therefore, the number of parts to be tested is 13.

Questions 2 and 3: The test has been performed and there is one failure among 13 parts. Calculate the confidence level [SIA 17] obtained to prove the reliability target and the number of new parts to be tested, without failure, to prove the reliability target?

$$Objective = Beta(F(x), x + 1, N - k + 1)$$

Excel ®: Confidence Level = BETA.LAW.N (WEIBULL.LAW.N (115,550; 3.3; 228,500; TRUE); 115,550 + 1; 13 – 1 + 1; TRUE) = 41.5%

$\delta = F(x) \geq 0.1$ for a well-defined test

Therefore, the confidence is 41.5% instead of 75%; if there is one failing part out of the 13 tested. The test is therefore inconclusive.

What is the number of new parts to be tested without failure in order to prove the reliability target?

Thirteen new parts are needed to meet the 75% confidence level. There is no other choice than to redo the complete test.

B) Illustration of zero-failure test with different results

Let us assume that you have received three different proposals:

– Supplier A: 150,000 activations, 5 pieces

– Supplier B: 120,000 activations, 10 pieces

– Supplier C: 100,000 activations, 22 pieces

Questions 4 and 5: What is the confidence level of the test to prove the reliability objective: for each of the suppliers? Which supplier provides the most appropriate reliability test?

$$c = Beta(F(x); x + 1; N - k + 1)$$

Excel ®: Confidence = BETA.LAW.N (WEIBULL.LAW.N (150,000; 3.3; 228,500; TRUE); 150,000 + 1; 5 + 1; TRUE)

Response to Q4: The confidence is:

– Supplier A = 77.60;

– Supplier B = 73.11%;

– Supplier C = 77.79%.

Which supplier provides the most suitable durability test?

Response Q5: Supplier A provides the best reliability test. The confidence in supplier B's test is too low. The confidence in supplier C is sufficient, but the test is too short: delta = 6.3% (delta = 10% is the minimum value).

Recall, $\delta = F(x)$ must be ≥ 0.

Recall, $\delta = F(x)$ must be ≥ 0.1 for a valid test, 1 for a valid test.

Illustration of a failed test

The test was done with seven parts. We have received the results (see Table 1.13): each piece failed after *N* activations.

N activations to failure
187000
221000
235000
260000
320000
357000
437000

Table 1.13. *Failure test results*

Question 6: What are the values for beta and eta – Weibull distribution parameters – for a 75% level of confidence?

We analyze these failure test results using Johnson's median rank method.

First, arrange the results of the *N* activations in ascending order.

Calculate the table below:

Excel ®:
Fi = SI ($ ri <> ""; INVERSE.BETA (confidence; ri; nb_parts-ri + 1); "")
Ln (x-d) = SI ($ ri <> ""; LN (zi); "")
Ln Ln (1/1 (1 – F)) = = SI (Fi <> ""; LN (LN (1/(1 – Fi))); "")
Slope = SLOPE (Ln (x – d); lnln (1/(1 – F)) = 3.13 = beta
Y-intercept = ORIGINAL.ORDER (Ln (x – d); lnln (1/(1 – F)) = –39.28
Quality of Model R^2 = (CORRELATION.COEFFICIENT (Ln (x – d); Lnln (1/(1 – F))^2

Eta = EXP(–Slope/order) = 287,140

See the good quality of the regression with the R^2 value.

Question 7: What is the reliability obtained? Are we within the target?

$$P_f = \int_0^{+\infty} F_R(z, \theta_{R1}, \theta_{R2}) \times f_C(z, x\ km \text{ or } y \text{ years}, \theta_{C1}, \theta_{C2}) \times dz \leq P_f^*$$

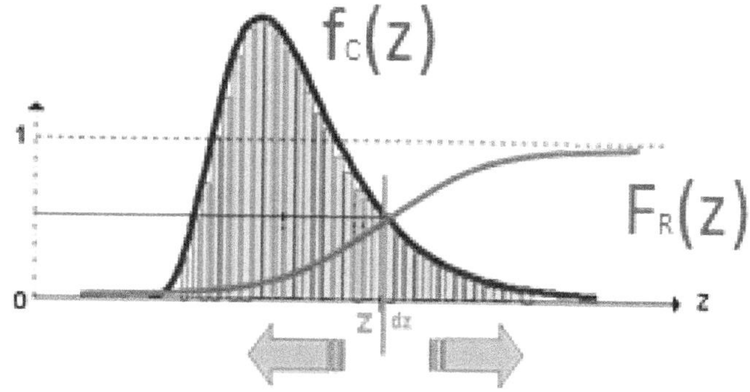

$Pf = 6.7 \times 10^{-5} < Pf^* = 10^{-4}$

The reliability target is met with a minimum confidence level of 75%.

1.3. Conclusion

This chapter describes a common method and illustrates this through a simple example. Reliability engineers size the parts that make up a vehicle with the view that keeps in mind product quality and the cost impact for the customer and the brand image. After the launch, they usually carry out a follow-up during production life through the analysis of the quality department's warranty returns. This verification ensures that the customer is genuinely satisfied.

1.4. References

[ABE 06] ABERNETHY R., *The New Weibull Handbook: Reliability and Statistical Analysis for Predicting Life, Safety, Supportability, Risk, Cost and Warranty Claims*, 2006.

[CCF 19] CCFA, "Durée de possession moyenne d'une voiture", *L'industrie automobile française*, available at: https://ccfa.fr/wp-content/uploads/2018/01/Analyse_Statistiques_2017_FR.pdf, 2019.

[DEL 16] DELAUX D., Warranty and reliability: new methodology for a good design, ESA AWARE, 2016.

[ISO 11a] ISO 26262 Véhicules routiers – Sécurité fonctionnelle Part 3, ISO26262-3: 2011, 2011.

[ISO 11b] ISO 26262 Véhicules routiers – Sécurité fonctionnelle Part 5, ISO 26262-5: 2011, 2011.

[PIN 79] PINNEL R., "Diffusion-related behaviour of gold in thin film systems", *Gold Bulletin*, vol. 12, pp. 62–71, 1979.

[SIA 17] SIA, Glossaire de langage commun fiabilité, available at: https://www.sia.fr/publications/432-glossaire-langage-commun-fiabilite, 2017.

[SIA 19] SIA, Guide d'aide à l'estimation et à la validation de la fiabilité automobile, Référence: DC-01-03, available at: Société des ingénieurs de l'automobile: https://www.sia.fr/publications/431-guide-aide-estimation-validation-fiabilite-automobile, 2019.

2

Structural Diagrams to Validate the Reliability of Mechanical Components

2.1. Introduction

When an accelerated test breaks a prototype, the design office sometimes challenges the result by saying that, *it is normal to break our prototype, given the excessive severity of the test. This does not make the component any less reliable for in service conditions*. The same conclusion would have been reached if the prototype had not broken during testing. Taking this reasoning a step further, accelerated testing could simply be eliminated, given that whatever the results, the conclusions are the same! This anecdote, which many reliability test managers could tell, highlights the major challenge in developing accelerated testing: knowing how to credibly link its results to in-service reliability. If we can demonstrate that a given result leads to a given level of in-service reliability for the component that will be manufactured at a large scale, we will have succeeded in developing a validation system that fully plays its role: filtering out unreliability and giving a timely warning when future reliability in series production is insufficient.

The credibility of an accelerated test can be quantified by its confidence level γ, defined by its probability of detecting a level of reliability deemed unacceptable in service.

Let us assume that the unacceptable level is set to $F < 1\%$. F is the probability that a failure will occur before a given service life t. Not generating more than 1% failure is a very unambitious goal for many components. Under normal use, a

For a color version of all the figures in this chapter, see http://www.iste.co.uk/elhami/applied3.zip.

Chapter written by Paul SCHIMMERLING.

probability calculation shows that 230 tests and no failures must be performed in order to demonstrate that this objective has been reached with a confidence level of $\gamma = 90\%$.

In fact, γ is the probability that the test will trigger a warning, that is the probability of occurrence of at least 1 failure among n tests. When the probability of failure is F, the confidence level can be evaluated with the binomial distribution:

$$\gamma = 1 - (1 - F)^n \qquad [2.1]$$

By performing $n = 230$ tests, we have $\gamma = 0.9$, when $F = 1\%$. This relationship shows that accelerated testing is essential to demonstrate high levels of reliability. By increasing the applied stresses, as compared to their in-service level, we increase the failure probability F, which allows us to reduce the number of tests n. However, it is important to know how to relate the results of accelerated testing to in-service reliability.

Experience in the validation of automotive mechanical parts has allowed us to identify some methodological diagram tools through which to establish this link. These diagram tools allow us to specify the hypotheses and the data to be collected in order to design and size credible validation tests. Designed for the validation of automotive components, these diagram tools can be applied to the validation of any mechanical part intended for mass production, subject to a well-known and identified critical failure mode. The sizing consists of determining the test profile (number of cycles to be applied, amplitude of the cycles, test duration, etc.), the number of parts to be tested and the acceptance criterion of the test so as to demonstrate the achievement of the specified reliability objective. The approach is also applicable to the definition of loading data for a numerical simulation that faithfully reproduces the identified critical mode. We will illustrate the approach through the development of a reliability validation test for disc brake pads.

2.2. Choice of methods

2.2.1. *Criteria selection*

Two criteria appear to be decisive. Firstly, the existence of a technical reference in terms of customers presenting the same failure modes as the prototype, and secondly, the nature of the failure: is it sudden or is the degradation progressive?

When a technical reference has already been put into service, it will be possible to exploit feedback. Feedback includes all the information reported by after-sales services. In the automotive industry, this means that the repairs carried out by the dealer and garage network. Feedback can also refer to failures and incidents, as well as examinations and measurements of used, non-faulty parts. The absence of incidents observed on a volume of parts in service also constitutes statistically exploitable feedback. The previous formula, equation [2.1], shows that the probability of failure is less than F, with a probability of γ, if the n parts put into service do not fail.

In the absence of a technical reference – which is the case for innovations – the mission profiles will serve as the only information source that can be used to characterize the operating phase in service.

When the failure manifests itself as a progressive, not brutal, degradation, it will be very interesting to set up a measurement of it. We thus go from observations of the "broken, not broken" type, to a quantitative measurement. It will hence be possible to estimate an average degradation, instead of a proportion of failures, thereby reducing the number of tests necessary to restore a given level of reliability [SIA 17b].

2.2.2. *Four basic methods*

The combination of the above two criteria allowed us to define four basic methods for establishing the link [SCH 03, SCH 14, SIA 17a, SIA 17b]: the comparison of Weibull laws in service and under testing, the comparison of degradations in service and under testing, the stress–strength method and the stress–strength method with degradation analysis (see Figure 2.1).

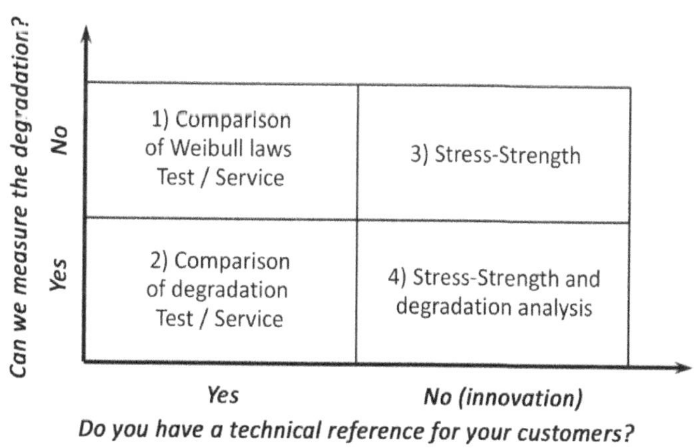

Figure 2.1. *Four basic methods linking reliability under testing to reliability in service*

2.2.3. *An applied example: the validation of disc brake pads*

The feasibility study of the four methods will be illustrated through the example of brake pads, presented in the guide [SIA 17b], which details the reliability calculations.

A hydraulic mechanism rubs the pads against a disc (see Figure 2.2). The critical failure mode, which is unavoidable, is abrasive wear-and-tear, which manifests itself as a loss of pad thickness, can be measured easily.

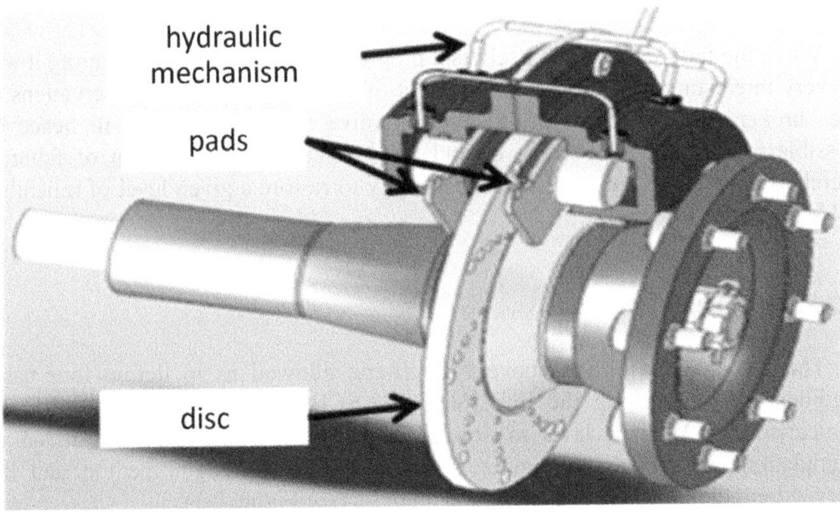

Figure 2.2. *Disc brake diagram*

A bench test simulates braking cycles by applying a force onto the piston of the hydraulic circuit's master cylinder in order to reproduce the effect of the brake pedal force.

A new brake will be fitted to a new vehicle. The reliability target set for the pads is $F < 1\%$ at 20,000 km. How can the test be sized to demonstrate that this objective is met?

The feasibility study will specify the data to be collected and the hypotheses to be consolidated in order to answer this question.

2.3. Feasibility study on the four methods

2.3.1. *Animation principle*

The reliability specialist will bring together at least one brake disc designer, one test manager and also a representative of the after-sales quality department. The principle is to review each of the four methods in order to specify with the working group, the data and the hypotheses necessary for their application. Each method will be explained through the aid of a diagram, so that all participants understand the principle and can give their opinion on the data and assumptions.

The facilitation principle is to ask questions in order to make the group clarify *the meaning of all the elements of the methodological diagrams* represented in Figures 2.3, 2.4, 2.6 and 2.7, in the context of the problem being studied.

Ideally, it is best to review all four methods, without focusing on one method a priori. Indeed, it is not always easy to discard a method a priori. The review of the four methods allows us to identify data that we may not have considered.

Before starting this analysis, we will first make sure that the following three prerequisites are met:

– B1: the test faithfully reproduces the failure mode observed in service. This leads to verify that the test activates the same damaging factors. The identification of the damaging factors *is facilitated by the identification of the physical mechanism of failure* at the origin of the material failure;

– B2: failure in service is caused by a physical mechanism, affecting all the parts in series and not by a poorly controlled manufacturing problem, affecting production batches (defects, non-conforming parts, etc.). The slope of the Weibulll laws on the after-sales data can help to identify this case [ABE 83, BLI 11];

– B3: a reliability objective to be reached in service is defined.

2.3.2. *Comparison of Weibull laws under testing and in service*

2.3.2.1. *Principle*

The method consists of comparing the Weibull law applied with the failures observed under testing and in service (see Figure 2.3). The test objective is readjusted if the serial reliability objective is not reached. The test target life (often specified by a Bq corresponding to $q\%$ failure) is multiplied by a factor K equal to the ratio of the series target Bq to the series Bq of the technical reference.

2.3.3. Comparing degradation under testing and in service

2.3.3.1. *Principle*

The idea is to measure the degradation for a sample of components, having the same technical definition for customers and for those under testing. The comparison of the degradations allows us to deduce the testing strategy with its acceptance criteria (see Figure 2.4).

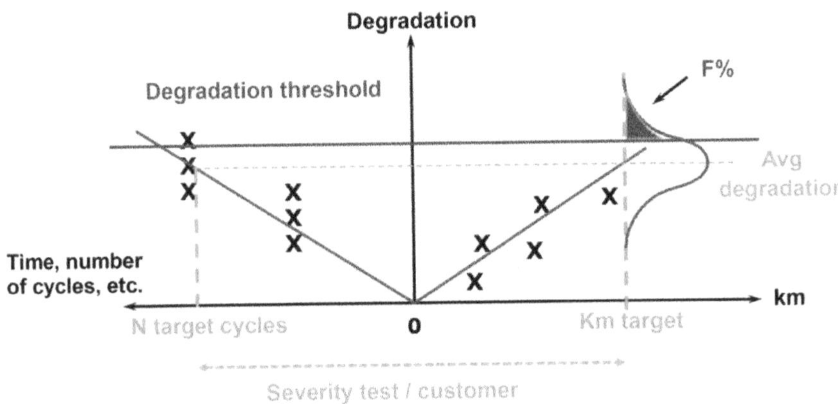

Figure 2.4. *Comparing degradation*

The severity of the test is obtained by identifying the test and service times that produce the same level of degradation.

The probability of failure F is determined by comparing a degradation distribution, at a given time, with a degradation limit. If this probability is too high, the average degradation not to be exceeded is determined, in order to respect the reliability objective and the corresponding test time.

2.3.3.2. *Applied conditions and assumptions*

The applied conditions and assumptions to be checked are:

– C1: know how to measure a degradation criterion that is representative of the failure in service;

– C2: it has joint measurements of degradation, in service and under testing for the same component. The sample of in service measurements must be representative of the whole population, which thus requires random sampling. The number of in service and testing measurements must be sufficient in order to estimate the degradation dispersion with an acceptable degree of uncertainty. A minimum of about 30 measurements is recommended;

– C3: this defines a degradation threshold, above which the failure is perceptible;

– H1: a law defining how the average degradation evolves over time must be postulated. Such laws have been proposed for elementary physical degradation mechanisms, such as wear-and-tear, corrosion and fatigue cracking [ARC 53, BEL 89, BAT 97, FLA 14]. When the crack initiation phase is preponderant when compared with that of the propagation phase, which is very fast, the fatigue failure is brutal. In this case, a progressive degradation model cannot be identified;

– H2: a statistical model must also describe how the dispersion of the degradation, quantified by a standard deviation σ, evolves over time. Several models have been proposed. A trajectory model, shown in Figure 2.5, leads to $\sigma(t)$ being proportional to t. A linear regression model will assume that $\sigma(t)$ is constant. Stochastic processes like Gamma or Wiener leads to $\sigma(t)$ being proportional to $\sqrt{t}$ [BAU 11, GRA 18].

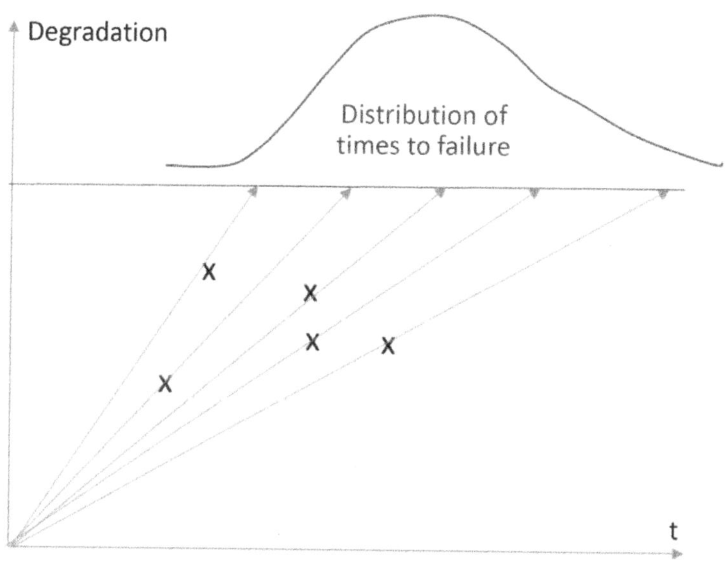

Figure 2.5. *Extrapolation of times to failure through a trajectory model*

2.3.3.3. *Variation: exploiting a Weibull law modeling in-service reliability*

Suppose that a Weibull distribution is applied to data from incidents that took place in service. The Weibull distribution allows us to estimate F at different times.

We can then determine the evolution law for the average degradation as a function of time by matching as closely as possible the probability of F as estimated by the Weibull law and F as estimated by a degradation model. This variant saves a sampling campaign for parts in service.

2.3.3.4. *Applicability to brake pad validation*

Condition C1 leads to the implementation of a method to measure a loss in pad thickness. Condition C2 launches a campaign to recover pads in the network. Condition C3 leads to define the thickness of the pad as the degradation limit.

The hypothesis H1 is taken from Archard's law [ARC 53]: the loss of thickness is supposed to depend on the mileage or the number of brake strokes through a linear function, to be adjusted according to the wear-and-tear measurements, once these are available. The statistical analysis of this data will also allow for determining how the standard deviation of the wear-and-tear evolves and to choose the H2 hypothesis.

2.3.4. *Stress–strength method*

2.3.4.1. *Principle*

The basis of the stress–strength method, described in several texts [SIA 17a, GOM 00, SCH 03, MAR 74], consists of defining a failure condition in the form of:

$$C > R \qquad [2.2]$$

C designates a quantifiable constraint designating the stresses, in the broad sense, undergone by the component in service: mechanical stresses, but also the number of solicitations, duration, number of operating cycles, etc. [MAR 74, GOM 00]. The mission profiles in service allow us to characterize the distribution of C, which is a random variable.

R is the resistance strength of the component to stresses, which is characterized by the test means. The results of the validation test determine the statistical distribution of R (see Figure 2.6).

The probability of failure F is estimated by $F = \text{Prob}\ (C > R)$.

The sizing of the reliability test consists of verifying that the resistance strength R is high enough to guarantee that $F < F_{obj}$ [PON 04a].

as data to characterize the distribution of C in service, that is, the les associated with each damaging factor. It will also be necessary to ndependence and correlations between damaging factors;

has data characterizing the distribution of the resistance strength, and larly the shape of the resistance strength law.

riant: case study on low variable uses

e service conditions are not very variable, C becomes a deterministic s case, the approach is to identify an accelerated life test (ALT) model s shown in Figure 2.7.

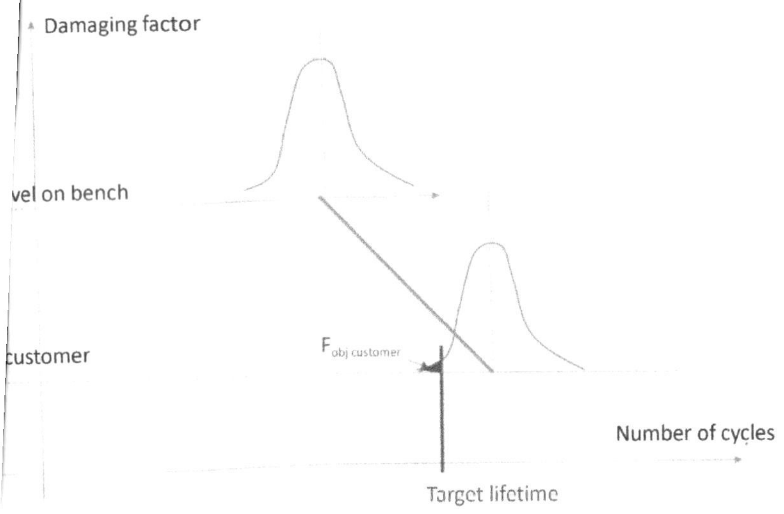

Figure 2.7. *Accelerated life testing model*

acceleration model allows the prediction of the service life distribution normal conditions, and the reliability in service through the results of ated tests. Different acceleration laws are recommended for different physical mechanisms: fatigue, corrosion, wear-and-tear, relaxation, creep [NEL 04, 3, BEL 89, BAT 97, FLA 14]. Their determination becomes easier as the parameters of the service life distributions, the slopes linking the damage to the service lives, become capitalized. SIA [SIA 17d] illustrates this ization approach for fatigue.

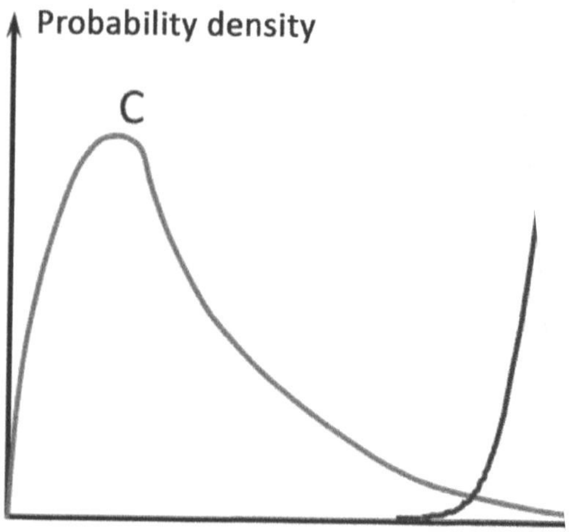

Figure 2.6. *Stress–strength*

COMMENT.–

The estimate for F is very sensitive to the shape an and R. It is prudent to re-calibrate the model with resp it is recommended to apply the method to components known, and to verify that the predicted reliability is reliability [BAS 18].

The method is applicable for innovative or non-innov

It is adapted according to considerations of variable u of automobiles.

2.3.4.2. *Application conditions and assumptions*

The conditions to check are as follows:

– C1: it defines a failure condition in the form of *C* comparable to *R*;

– C2: it defines a synthetic damaging factor, representin several damaging factors [PON 05]. This requires knowing each influencing damaging factor;

2.3.4.4. *Applicability to brake pad validation*

The failure condition C1 can be defined in several ways.

First solution: C is the number of brake applications applied at the 20,000 km horizon. R is the number of brake applications at failure and observed on the bench test.

A criticism can be made about this formulation. The brake applications in service do not all have the same intensity as the brake applications for the bench test. Hence a second solution: define a synthetic damaging factor, combining the number of brake strokes, the maximum pedal force and its duration of application. This work must be done with braking specialists. Archard [ARC 53] proposed a wear-and-tear law which allows for a preliminary approximation.

$$U = k.\frac{Q\,d}{H} \qquad [2.3]$$

where:

- U: usage volume;
- k: wear-and-tear coefficient;
- Q: pressure applied to the contact;
- d: distance traveled;
- H: hardness of the softest body.

The distance covered d under abrasion is proportional to the number of brake strokes N and the brake stroke duration D.

Equation [2.3] shows that the same usage volume U can be obtained by applying N_{eq} brake strokes to the bench with an effort of Q_{bench}, for a duration D_{bench}, N_{eq} verifying:

$$U = (k/H)\cdot Q_{bench}\cdot D_{bench}\cdot N_{eq} = (\mathrm{k/H})\cdot N\cdot D\cdot Q$$

where: $N_{eq} = N\,D\,Q/(D_{bench}\,Q_{bench})$ [2.4]

The application condition C3 leads to the construction of the distribution of N_{eq} from available records of $N\,D$ and Q.

A third solution consists of retaining as the condition:

cumulative wear-and-tear > pad thickness

[GRA 18] GRALL A., "Modélisation d'une cinétique de dégradation", in LANNOY A., LEMAIRE M., DELAGE A. (eds), *La Fiabilité en mécanique,* Presses des Mines, 2018.

[MAR 74] MARCOVICI C., LIGERON J.-C., *Utilisation des techniques de fiabilité en mécanique*, Lavoisier, Techniques et Documentation, 1974.

[NEL 04] NELSON W.B., *Applied Life Data Analysis*, Wiley, 2004.

[PON 04a] PONCET A., SCHIMMERLING P., JOUZEAU C., et al., "Le sizenement des essais de validation par la méthode Contrainte-Résistance", *Actes du Congrès Lambda Mu 12,* Bourges, 2004.

[PON 04b] PONCET A., SCHIMMERLING P., MIAZGA D., et al., "Sizenement de chapeaux de paliers de moteurs turbo diesel par une approche fiabiliste", *Conférences SF2M,* Paris, 2004.

[PON 05] PONCET A., Fiabilité expérimentale. Sizenement optimale d'essais sous contraintes Modélisation des facteurs endommageant le composant et intégration dans le sizenement de l'essai, Application à l'automobile, ENISE, 2005.

[PRI 05] PRINCE O., MORIN G., JOUZEAU C., "Validation test optimization based on a statistical approach for diesel engine cylinder heads reliability", *SAE Conference,* Detroit, 2005.

[REL 18] RELIASOFT, *Weibull++ Alta Users's Guide*, available at: http://www.synthesisplatform.net/WeibullALTA/en/UG_WeibullALTA18.pdf, 2018.

[SCH 03] SCHIMMERLING P., PRINCE O., "Prévoir la fiabilité en clientèle à partir de données du réseau, de résultats d'essais ou de calculs. Quatre méthodes de bases illustrées par des exemples", *SIA conference "Exploitation des données du réseau pour estimer et maîtrise la fiabilité*", Paris, 2003.

[SCH 05] SCHIMMERLING P., PRINCE O., JOUZEAU C., et al., "Sizenement des plans de validation par une approche Contrainte Résistance combinée à des mesures de dégradation. Application à la validation des culasses et des injecteurs", *Conférence S.I.A "Conception des Plans de Validation",* Paris, 2005.

[SCH 14] SCHIMMERLING P., NDIAYE H., CASTRIC S., et al., "Formalising industrial reliability problems", *Conference ALT,* Pau, 2014.

[SIA 17a] SIA, Guide d'aide à l'estimation et la validation de la fiabilité automobile, version 3 Document collaboratif SIA, DC 01 03, Rubrique Publications, Suresnes, available at: S.I.A www.sia.fr, 2017.

[SIA 17b] SIA, Handbook for Automotive Reliability, Assessment and validation, DC 04 01, Rubrique Publications, Suresnes, available at: www.sia.fr, 2017.

[SIA 17c] SIA, Glossaire de langage commun de fiabilité, Document collaboratif SIA DC 02 02, Rubrique Publications, Suresnes, available at: www.sia.fr, 2017.

[SIA 17d] SIA, Préconisations pour les caractéristiques statistiques de résistance en fatigue. Application aux aciers et autres matériaux utilisés dans la construction automobile, Document collaboratif SIA DC 05 01, Rubrique Publications, Suresnes, available at: www.sia.fr, 2017.

3

How to Put an Efficient Methodology to Design Innovative Products in Place

3.1. Introduction

The evolution of the world economic market necessitates a reorganization of all economic sectors, in particular, the sector for new technologies. The growing competition, the innovation of processes and procedures, the increase in the speed of data transmission, the increase in the mass of data to be processed (Big Data) whose exploitation is exploding because of the evolution of the processors, the needs of the customers, the protection of the data in the face of external threats, etc., all of these necessitate a reliable and robust approach for the product development process.

3.1.1. *Reliability*

The *Union Technique de l'Electricité*[1] (UTE), on the recommendation of the International Electrotechnical Commission, has proposed the following definition: Reliability is the ability of a device to perform a required function under given conditions for a specified period of time. Reliability is the probability of not having any failure during this period of time *t*. It ranges from 0 to 1 (or 0–100%) and is denoted as *R(t)* (*R* for reliability and *t* for time). Reliability, as a function of time, should not be confused with quality control (static function).

For a color version of all the figures in this chapter, see http://www.iste.co.uk/elhami/applied3.zip.

Chapter written by Claire SCHAYES, Ludovic NGAVOUKA and Eric MANOUVRIER.

1 UTE is the French national organization for normalization in the domain of electronics. It is a member of the International Electrotechnical Commission and of the CENELEC.

3.1.2. *Variability*

Variability is the tendency for a process to produce slightly different results or measurements on the same test item, where the measurement conditions are either stable or vary over time depending on the environment, the operators and sometimes the *analysis method* used. The case study we will discuss considers two sources of variability:

– instantaneous variability in the short term, ensuring that the precision of the measuring device is adapted to the measure;

– long-term time-dependent variability, taking into account modifications due to environment, manipulations and the various interventions that change the process.

Continuous improvement is one of the essential approaches allowing for regular and constant follow-up of a quality-control system for a process. W. Edwards Deming is one of the founders of the concept of continuous improvement, which is the first objective of the quality control system. The Deming PDCA (Plan-Do-Check-Act) cycle, which is one of the tools for continuous improvement, illustrates how to achieve continuous improvement for any process. By extension, "*Lean Six Sigma*" is now widely used across the industry as a methodology for managing and organizing processes.

3.1.3. *"Lean Six Sigma"*

"*Lean Six Sigma*" is a quality-control method initiated by Motorola and then used at General Electric (Jack Welch), Texas Instruments, Honeywell, IBM, etc. It is a centralized approach with a strong culture of *continuous improvement*, focusing on customer needs, products and processes; it is essential to emphasize the analysis of critical and essential customer requirements in order to avoid any disruption or variation between the different process steps or entities within an organization (customers, projects, R&D, processes, production, quality, etc.).

It allows us to set up a real culture of continuous improvement within an organization by simultaneously working on physical and information flows because of two complementary approaches:

– The *lean* approach aims for the identification and elimination of losses (MUDA) without any added value for the customer, the company and/or the employees, as well as aiming to decrease the rigidities of the company in order to be adapted as close as possible to the demand. Lean allows us to value the competence and expertise of everyone, especially in the cases of acquisitions and mergers of new companies within an entity.

– *Six Sigma* aims to eliminate defects and reduce variability at each stage of a process so as to better satisfy customer needs. It promotes a scientific culture within operations management. The "Six Sigma" is a statistical approach that allows us to select, compare and improve a system, a subsystem and its components for a better integration, optimization and verification of the whole system's capabilities.

3.1.4. *Quality according to the "Lean Six Sigma" approach "is conforming to requirements" [SMI 14]*

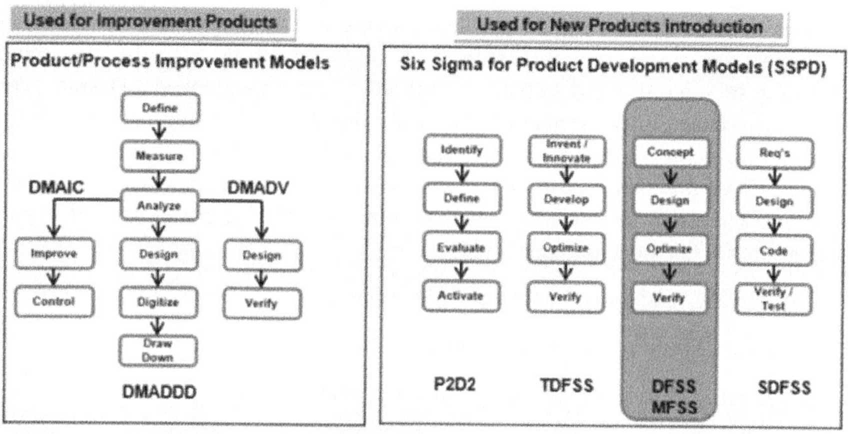

→ **All SSPD** models are used for developing new products:
- **P2D2 :** Product Portfolio Definition and Development,
- **TDFSS :** Technology Development for Six Sigma,
- **DFSS :** Design for Six Sigma,
- **MFSS :** Marketing for Six Sigma,
- **SDFSS :** Software Design for Six Sigma.

Figure 3.1. *Diagram showing how DMAIC, DMADD, DMADV and DFSS correspond to different phases*

These projects revolve around two types of strategies:

1) the model used for *improving* projects or processes;

2) the model used for *innovation and design* projects and the introduction of new products or processes:

- for improvements, some organizations use DMAIC, DMADD (especially in digital data management) and DMADV (for improvement of an existing product or process),

- for the introduction of new products, Design For Six Sigma (DFSS) is required.

3.2. DFSS

DFSS is used for innovation, and design projects enable a customer-oriented product design process; it has been presented as an approach to product development that complements the Six Sigma problem solving methodology. This methodology encourages the systematic anticipation of customer needs and the disciplined application of scientific and statistical methods to meet those needs. Despite these naming differences, all versions of DFSS share fundamental strategies and tools that promote a common goal: *to create a culture of product development that is based on customer needs and meets customer requirements for experimental and operational reliability*. The important deliverables that comprise DFSS are summarized in each phase of the DCOV development cycle:

– D: activities and tools for the definition phase;

– C: activities and tools for the characterization phase;

– O: activities and tools for the optimization phase;

– V: activities and tools for the verification phase.

a) DFSS and experimental reliability

In Lean Six Sigma, experimental reliability studies allow us to identify the most fragile components of a system, and the key performance parameters are defined and monitored to anticipate and reduce the risks of variation; the main elements are selected to make statistical correlations.

It involves implementing product or *process design initiatives* in a development methodology that relies on DCOV process development phases. It is also used to apply statistical analysis and simulation tools to identify the *sources of variation* of the system and its components in order to set up models that allow for reliability analysis throughout the process.

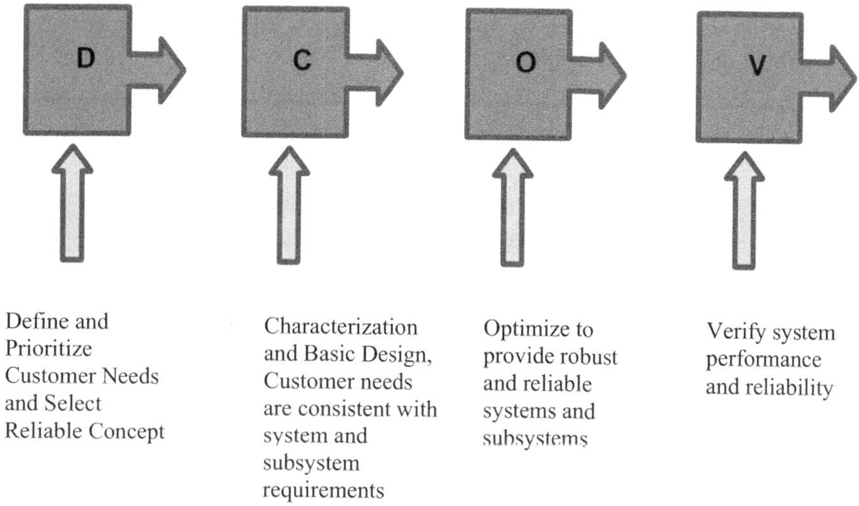

Figure 3.2. *The identification and analysis of the sources of variability begins at the definition phase and continues until the verification phase*

It is a method that allows us to translate the customer's requirements into technical requirements at each phase of the product realization cycle; the system level is the starting point of the analysis. This only makes sense if there is a required continuity in the form of deliverables between early-stage projects, innovation projects, customer projects and productivity projects. *This continuity is today a real difficulty in all large groups that are constantly trying to find the best sequence of performance tools, the "famous" process approach.*

b) DFSS and operational reliability

The reliability of electronic systems depends on several factors, but the method or approach selected to ensure and guarantee this reliability remains the key to success. This method must take into account all of the development tools, the internal and external qualification and validation processes, and the company's stakeholders, also taking into account the interactions between each development phase. For example, Table 3.1 illustrates how the Ford Motor company implemented Lean Six Sigma within their organization, whereby the company promoted the philosophy "Quality is job number one" in the 1980s. Today, they have demonstrated that making product quality a top priority is more than a slogan; it is essential to the company's operations and long-term growth, to implement a methodology [PIL 17].

Table 1. Ford Technical Education Program

Ford technical education program subjects Web based, self-paced training with tests	
Process fundamentals: DMAIC, DCOV*	Reliability
System engineering fundamentals	Statistical engineering
Applied consumer focus	Experimental design
Design verification and production validation	Parameter design
Failure mode and effects analysis	Tolerance design
Eight discipline (8D) problem solving	Process control methods

* Added in 2004; all other subjects were available prior to DFSS deployment.
DMAIC = define, measure, analyze, improve, control.
DCOV = define, characterize, optimize, verify.

Table 3.1. *Ford technical education program*

The automotive industry was an early adopter of Lean Six Sigma and continues to lead the way in showing other industries the difference Lean Six Sigma can make. Companies like Toyota and Ford have shown that Six Sigma can dramatically reduce costs and non-quality defects, freeing up financial resources for other areas of the business. They have shown that improved quality saves money. Fewer scrapped products, fewer warranty repairs and replacements, and much higher levels of customer satisfaction and retention resulted in increased profits for both companies [PIL 17].

c) Performance and applied reliability

Key performance indicators (KPI) of a company are simultaneously a tool to gauge the general health of the company and a decision-making tool. They affect all areas of activity for the company: they allow us to determine production efficiency.

The concept of performance first integrates the notion of efficiency, that is, the idea of undertaking and completing an action. Performance therefore consists of obtaining a certain result in accordance with a given objective. Within a company, the objective must be:

- consistent with the company's purpose;
- previously defined and measurable;
- accompanied by the expected result.

The objective need not be quantified, but the result must be measurable.

d) The concept of efficiency

The concept of performance thus integrates the notion of efficiency, that is, the idea that the means used to carry out an action have been exploited with a view to economy. An action will therefore be considered efficient if it allows the expected results to be obtained at the lowest cost. Performance is an optimal result obtained through the most efficient use of resources. Performance is therefore always a relative concept.

In summary, performance refers to the ability to carry out an action in order to obtain results, in accordance with previously set objectives, all the while minimizing (or making profitable) the cost of the resources and processes being implemented.

e) Interactive practice

We propose to immerse the reader in a concrete case study, in the form of augmented content to be downloaded, which will be commented on by resuming the DMAIC sequence previously introduced. Our analysis will be process oriented. For this interactive experience, the reader will first connect to the following link: *here* (to be hosted on the editor's website) (or QR code?).

Instructions: Download, unzip, open "Story.html or Story.html5" to launch the course. You will be able to download a training module which details, step by step, a variability study within the framework of an improvement project, conducted according to the DMAIC method. We advise you to complete this module now and then return here in order to take full advantage of the rest of this chapter, which is its extension.

3.3. DMAIC

3.3.1. *Introduction to DMAIC*

DMAIC is probably the best-known method in the industrial world since it is the subject of the NF X06-091 Standard of April 2011 [NOR 11], which specifies the required skills and deliverables to achieve Lean, Six Sigma and Lean Six Sigma projects. The DMAIC project is broken down into five phases: define, measure, analyze, innovate and control.

The objective of this book is to ask ourselves: how does DMAIC allow us to control the variability of industrial processes? Most of these Retex[2] have never been published since they are based on the practical experience of authors and illustrated through our case studies.

3.3.2. *Why launch DMAIC projects?*

3.3.2.1. *To reduce non-quality costs*

On average, non-quality costs can represent up to 2% of sales in the industrial sector. In most competitive sectors, the operating margin does not exceed 5% of sales, so it is easy to understand that non-quality costs can wipe out any profitability! In an ideal world, the principle of self-quality, a trait that is dear to Taïchi Ohno, the inventor of 7 MUDAS (waste), would mean that any upstream process would deliver a good part to the downstream process. In reality, each process is itself faced with important sources of variation (*Mura*) in terms of personnel (different teams and different training), the product (material, dimensions, physical characteristics, etc.), the environment (temperature, humidity, etc.), the organization (respect, efficiency, etc.), the tools used (digital, paper, etc.) and so on. The list is unlimited! The hunt for MUDAS is therefore not enough since the Mura represents a major source of non-quality, as well as the Muri. As a reminder:

The 3M (Muda, Mura and Muri) method identifies three enemies of lean manufacturing:

– Muda (waste): any activity in a process that does not provide value;

– Mura (irregularity): any variation leading to unbalanced situations;

2 RETurn of EXperience.

– Muri (excess): any activity involving unreasonable strain or effort on personnel, materials or equipment.

One of the difficulties is to align the internal and external process approach, especially when we buy primary parts from different suppliers located far from the assembly site. In such a situation, the principle of "*visual management*" (seeing the whole flow in one place) is difficult to apply in practice.

Let us take as an example: a plastic part whose supplier would give us a quality commitment of 3 sigma for a mechanical tolerance (in the sense of the ASQ[3] definition): MINITAB tells us that this means that 99.73% of the parts are in the brown section of this graph, and therefore, there are 1 – 99.73% = 2,700 parts per million outside of this zone.

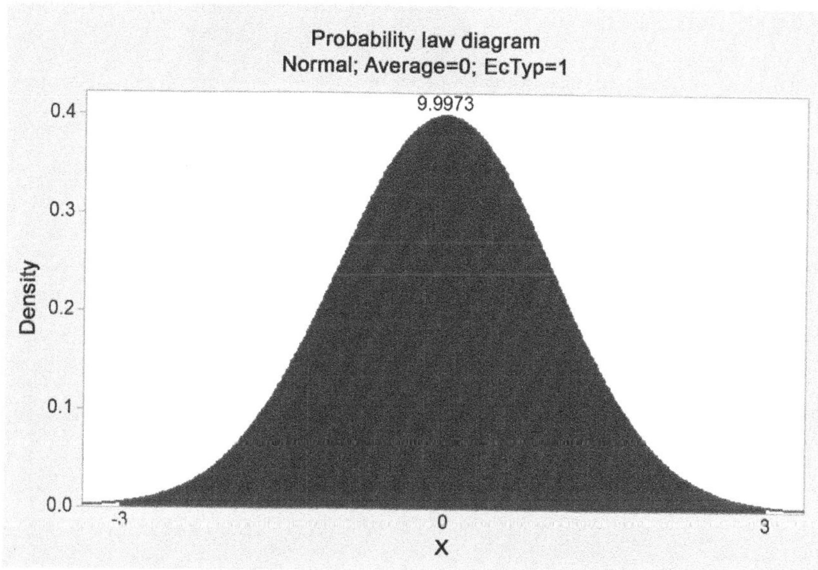

Figure 3.3. *Probability law diagram*

Now let us imagine a decentering of 1.5 sigma and look at how many parts would potentially end up outside the engagement zone.

The recalculation is simple 1 – 0.9332 = 66,800 ppm versus 2,700 ppm initially that is, a ratio of 24. The ASQ gives these two definitions: the one without drift (2,700 ppm at 3 sigma) and with drift (66,800 ppm).

3 American Society for Quality.

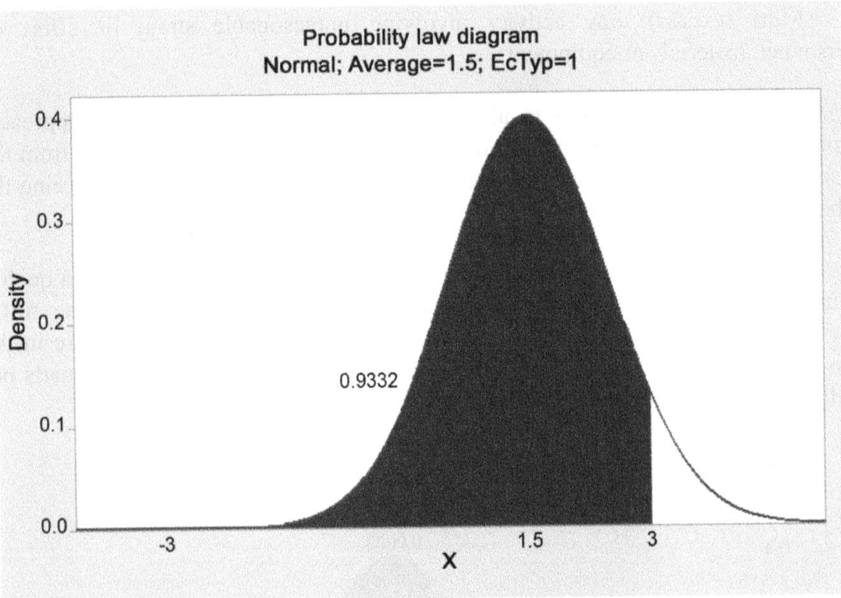

Figure 3.4. *Probability law diagram*

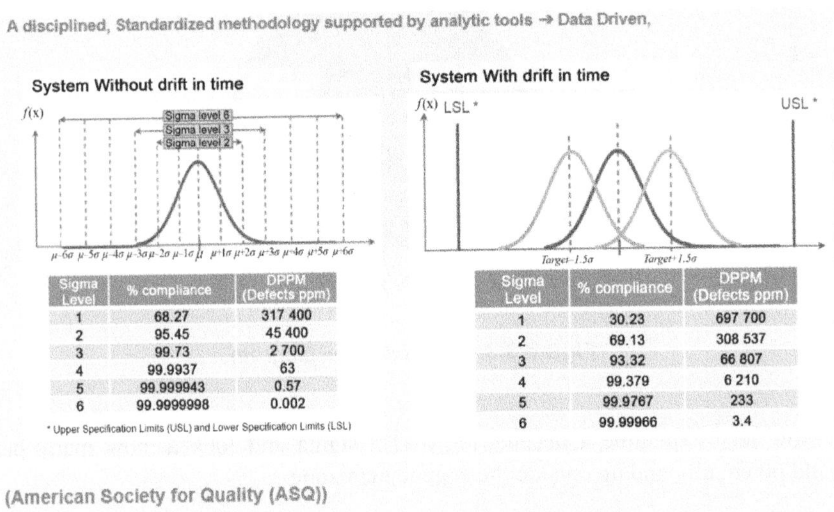

Sigma Level	% compliance	DPPM (Defects ppm)
1	68.27	317 400
2	95.45	45 400
3	99.73	2 700
4	99.9937	63
5	99.999943	0.57
6	99.9999998	0.002

Sigma Level	% compliance	DPPM (Defects ppm)
1	30.23	697 700
2	69.13	308 537
3	93.32	66 807
4	99.379	6 210
5	99.9767	233
6	99.99966	3.4

Figure 3.5. *A disciplined, standardized methodology supported by analytic tools → data driven*

Clearly, in order to control the expected level of quality, it is not enough to set limits at the outset, but rather ensure to control any drift over time. This is the notion of stability, which comes from the robustness of the industrial tool.

No study of variation can be done without measurement or without information being sent back to the servers. The DMAIC method is key to helping us stabilize industrial processes, either in the form of work sites or control at the start of lines (automatons or line concentrators). A lean specialist will always start by killing quality problems in order to make the flow more fluid, for which Six Sigma is an essential method. This is why the most reputable companies on the market use it as a strong foundation through which to ensure a continuous flow. One of the most widely read books on this subject is *Six Sigma* [PIL 13].

3.3.2.2. *How to reduce CAPEX (capitalized expenditure)*

Most financiers wonder about the impact of Lean Six Sigma on the profitability of budget requests. Therefore, F/V0 = IRR, where F is the cash flow brought by the machine, V0 is the market value and IRR is the internal rate of return. To increase the IRR, we must therefore increase F and decrease V0.

Here, we have shown that:

– decreasing the cost of the investment;

– increasing the cash flow generated (due to better design) increases IRR.

Remember that:

– in order to increase F, eliminating MUDAS (waste) is the best solution;

– in order to decrease V0: the LEAN method applied to design will have an impact in the order of 20–40% gains, based on the authors' experience. Moreover, one of the most telling books on this subject is "*Gold Mine*" [BAL 14], where the priority for a lean specialist will be to maximize flow while maximizing added value, thereby removing any source of waste and variations not paid by the final customer. The formula F/V0 = IRR shows that everything is done at the design stage, after which point it is too late or will need more human, financial and time resources, etc. Yet, 90% of companies only apply lean manufacturing a posteriori, once the machines are already on the ground and only 10% apply DFSS or preventive lean methods (a priori) to design, upstream of the machines that have already been optimized on the ground.

One of the clearest books on the subject is *Design for Six Sigma* [PIL 17], which is a good reference in terms of Lean Design, where a well-designed product does not generate Muda on the process. A *poka yoke* product will, for example, always be preferable to a *poka yoke* process. The authors regret that there is still no standard in this area of Lean Design (for either product or process) and insist on the importance of practicing a form of lean upstream, during the design phases, in order to control any subsequent variability once on the ground. It is understandable that it is difficult to impose a way of innovating across all companies, due to the diversity of business models; however, we can agree on the approach. It is obviously not possible to standardize a company or its unique method for innovation.

3.4. Feedback

3.4.1. *Feedback on the define phase*

Why launch DMAIC projects?

Phase D shows the importance of defining the need, the objective to be achieved. If you think this is obvious, simply answer this question: in your company, do all new product developments have a clear objective in terms of the cost to be achieved? A quantified quality objective? Deadlines to be met? (A clear project charter?)

In phase D, it is especially necessary to define a *multidisciplinary team* integrating all of the actors concerned by the global flow. Therefore, suppliers of machines and raw materials must be part of teams for the industrial sites concerned during the design phases. This team will meet frequently in the form of *workshops* (limited in time), and will have the opportunity to *brainstorm*. The authors recommend not exceeding *six people per group* for optimal efficiency. The authors' experience has shown that adding a seventh person reduces efficiency by half. Therefore, if the starting group is made up of 12 people, it should be separated into two subgroups for the brainstorming phases. If the starting group has 13 people, we will have to split it into three sub-groups!

3.4.2. *Feedback on the measure phase*

The question most often asked of a LEAN SIX SIGMA expert is how many trials to run in order to minimize costs, while still being able to make the most of the results. This type of matrix from MINITAB is a useful visual answer:

If we are looking to screen and thus not take into account the interactions between the factors, the resolution plans III and IV will be sufficient for a first approach. If we wish to study a more complete design, we will have to take into account the "COM" boxes. We will see a practical example applied to a welding machine later on.

Create a factorial design: View available designs ×

Available factorial designs (with resolution)

	Factors													
Trials	2	3	4	5	6	7	8	9	10	11	12	13	14	15
4	Com	III												
8		Com	IV	III	III	III								
16			Com	V	IV	IV	IV	III	III	III	III	III	III	III
32				Com	VI	IV	IV	IV	IV	IV	IV	IV	IV	IV
64					Com	VII	V	IV	IV	IV	IV	IV	IV	IV
128						Com	VIII	VI	V	V	IV	IV	IV	IV

Available Plackett-Burman Resolution III Designs

Figure 3.6. *Available factorial designs (with resolution)*

3.4.3. *Feedback on the analyze phase*

Many studies are interested in the variation of a factor on an output, for which we talk about sensitivity. However, industrial life will be more interested in the multi-variate study, which consists of studying the influence of several factors and/or their interactions between them on an output variable. For this purpose, we would like to indicate a second very useful matrix requested almost every week before launching any analysis, no matter the software being used.

	Outputs:	**Yi**	**Yi**
Influential factors:		**Continuous**	**Discrete**
Xi	**Continuous**	Simple or multiple regression	Binary or ordinal logistic regression
Xi	**Discrete**	ANOVA	Chi-square test

We insist that BIG DATA will explode:

– the so-called analysis of variance (ANOVA) studies, because many discrete Xi variables are now accessible with the development of digital and so-called MES systems;

– the so-called logistic regression studies with discrete Yi and continuous Xi. In Appendix 1 (see section 3.10), we give the example of an applied case study of ANOVA.

Variability studies can also be applied to cycle times. In Appendix 2 (see section 3.11), we explain how to approach this kind of study.

3.4.4. *Feedback on the innovation phase*

Based on the analysis reports, this phase must create a break between that which exists and the future. One of the most frequently used sources is to draw inspiration from a non-competing sector in order to find new ideas. A practical example in the automotive industry is to take inspiration from the "sliders" used in the pharmaceutical industry to bring aspirin tablets onto an encapsulation production line, without putting too much strain on them because otherwise they break. A slider only moves along one axis, the x-axis).

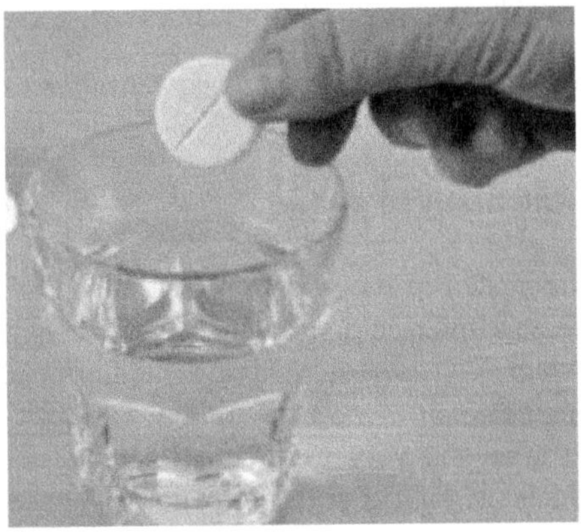

Figure 3.7. *Aspirin tablets*

However, often in the automotive industry, small thin parts have to be supplied and nine times out of 10, vibrating rails are proposed with a Y vibration, which can often damage very fragile parts, by making them overlap, causing blockages, causing them to break, etc.

One such lean project used the sliders employed for aspirin tablets in the automobile industry to eradicate quality concerns previously encountered on vibrating rails. This type of idea is common in Phase I. The best stimulus is expertise; it is essential to invite experts who will bring their pioneering ideas and backgrounds for the service of a new design. It is also important to go to industrial fairs, the largest one in Europe being the one in Hanover, where all of the latest technical innovations are presented (Cobots, Industry 4.0., digital tools, and all of the novelties in terms of Mechanical and Repaired Outlets). Even if industrial buzz is becoming commonplace, it is clear that the trend is to mix genres: between hardware and software; between Big Data and connected objects; between machines and servers; between virtual reality and existing systems. Another catalyst for creativity is to work through workshops and under time-bound projects.

For example, a formula of three days brainstorming and three days digital or cardboard mock-ups generates a creative explosion for the participating teams, because creativity is released more when time is limited than when cogitation is allowed to drag on. Finally, as already mentioned, given that it is impossible to know everything with the current speed of progress, it is therefore necessary to invite the MRO and machine suppliers concerned to participate in these workshops, because they will be the source of creativity external to the organizing company.

The eighth MUDA is moreover the non-consideration for the creativity of internal or external teams, and it is therefore advisable to eliminate this MUDA, as it can be particularly formidable during the innovation phase.

3.4.5. *Feedback on the control phase*

The control phase will obviously call upon process control charts, machine capabilities, the principle of auto-quality and Quick Response Quality Control (QRQC).

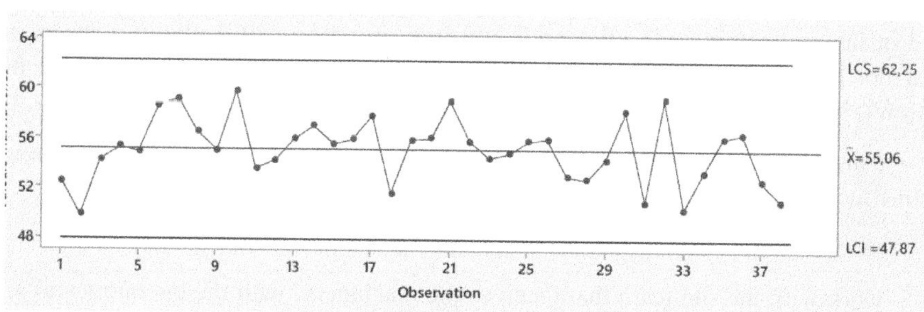

Figure 3.8. *Example of a control chart*

A Cp greater than 2 and a Cpk greater than 2 are excellent results. Concerning the QRQC, the authors refer to the book *Perfect QRQC* [AOU 12], which is the bible used at VALEO.

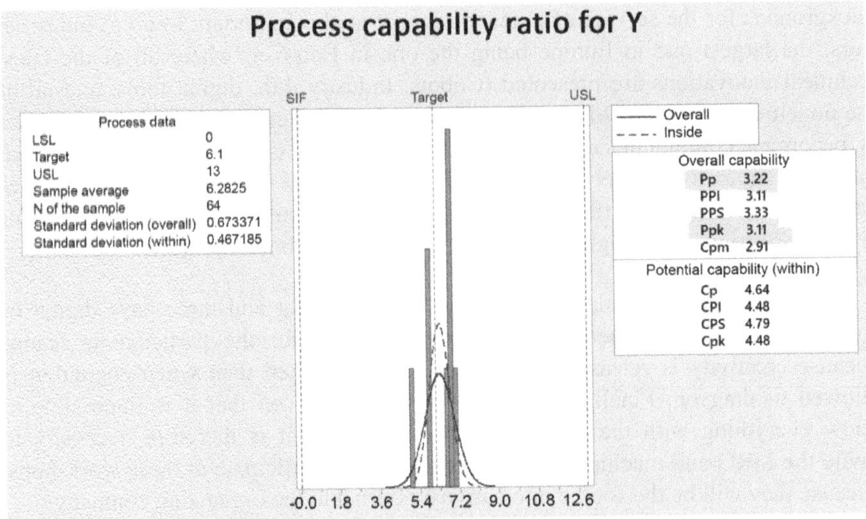

Figure 3.9. *Example of machine capability*

3.4.6. *Can DMAIC be customized?*

It is not up to the company to adapt to the Lean Six Sigma method, but rather for the method to adapt to the company, due to its specificities. Also, in some instances DMAIC is available under different forms, so as to be applicable to certain topics.

VALEO uses such a method to design its machines: it is the LEAN PROCESS DESIGN, or how to design a simple, competitive and robust machine, without waste and in line with safety and ergonomic requirements. For confidentiality reasons, it will not be possible to reveal this method here. However, the key points are as follows:

- consult suppliers on functional specifications in order to give them the free rein with which to express their creativity, and not to impose fixed technical solutions;

- set clear industrial objectives before starting to work as lean worksites;

- work with the site team that receives the machinery, with the operators and all those who will be involved during the acceptance phases at the supplier and on site;

- request a budget in line with the defined objective and launch the internal approval process for the duration of the worksites, so that the order can be launched at the end of the worksite;

– seek a technical breakthrough (40%) at each site in terms of the most relevant industrial indicators: surface, quality, KOSU[4], cost, etc.;

– systematically produce a scale model to validate all the risks taken, in digital form or in the form of cards;

– systematically carry out MUDAS eliminations in the field; the basis of the Toyota method by Taichi Ohno in the 1970s. If the process does not yet exist, do it on a machine that is similar, although this phase is mandatory and must be done where the machine can be physically affected;

– do not hesitate to study all possible assembly sequences of the *Ishikawas* on the floor with the real components in hand;

– always take into account the facts brought up; let the evidence speak for itself;

– invite the right experts, from the right fields of expertise for the subject being considered;

– invite N-2 rank suppliers, if necessary, for key topics such as vision, robotics, testing, etc.;

– always start by working on the assembly sequence, then check that the corresponding flow (with material supply) is coherent. Next, move onto the study of cycle times. The three key questions are thus always in this order: What is the sequence? What is the flow? What are the operations?

– validate all of the hypotheses considered before launching the studies, in the form of prototypes or digital simulations;

– capitalize the processes that have expected results on the Gemba[5] as standards;

– use only the Lean tools that are relevant to the subject concerned, in order to be efficient, and do not hesitate to invite specialists for the right tools if necessary;

challenge the "in-house" method by a permanent external questioning, thanks to the various existing Lean networks via the *Grandes écoles*, the *Grandes*

4 KOSU is a Japanese word. It is a measure of productivity. It refers to the number of man hours it takes to produce one unit of a product, or to complete a process. It is calculated by multiplying the number of workers involved in making a product by the number of hours used. That gives the total man hours. That number is then divided by the total number of units produced during that time.

5 Gemba (also written as genba) is a Japanese word meaning "the actual place". In lean practices, the gemba refers to "the place where value is created", such as the shop floor in manufacturing, the operating room in a hospital, the job site on a construction project, the kitchen of a restaurant and the workstation of a software programmer.

entreprises and Lean organizations, such as the Lean Six Sigma University [LEA XX], the Lean France Institute [INS XX] or the excellent WikiLean website [WIK XX]. In a word, each company practices, without naming it as such, a DMAIC approach, which is available for the appropriate phase, according to the specificity of the subject. An exchange is always useful between companies from different, non-competing sectors. Recently, an exchange took place between the VALEO Group and the ELECTROLUX Group.

Bernard Demol, Head of Global Supplier Development, explains:

> As Head of Supplier Development, my objective is to improve the manufacturing processes of our strategic suppliers in terms of flexibility, capacity and performance quality. Quality is a way to touch the product through a 360° approach. That is to say, through R&D, for example: how to integrate product poka-yoke as early as possible, through quality, by challenging the current status quo against future performance levels, through automated production, for example, and through logistics between transport optimization versus improving packaging definition and its reuse. However, all these improvements are possible, provided the environment is conducive to exchange and has the capacity to be questioned. These are win-win scenarios where everyone finds their place and their interest. Lean Six Sigma is one of the ways by which to achieve a climate of transparency, through the use of data and participation, and through improvement projects involving different functions (from the operator to the methods engineer).
>
> I had the pleasure to welcome VALEO during one of my seminars attended by my teams from across the world. On this occasion, we wanted to exchange and share our practices and experiences in terms of involving suppliers in the context of new products and therefore new investments. What we learned, and try to apply every day, is that improvement is infinite and that there are always ways to go beyond preconceived ideas or the current state of things. That there are many more ideas generated by a group of people than by one person. And finally, that we should not aim for perfection but try, learn, and improve: failure is part of learning!

3.5. How to design a reliable welding process with control over the design of experience?

Design of experiments (DOEs) are used in different domains such as the chemical formulation of new cosmetic products [MAT 00], recipe creations within the food industry [SAB 06], but also the optimization of industrial process, such as soldering or welding [BID 16]. We wonder how DOEs may be used for such varied applications. In fact, in all the examples cited, there is always an objective to be achieved related to the process, for example, a cream that nourishes the skin without causing allergies, a pizza dough that cooks and crisps in 10 minutes, a soldered connection that has a nice visual appearance and whose mechanical strength is that required for its intended application. In all cases, there are also variables, called factors, for example, the composition of the materials to be joined (copper, steel, aluminum, etc.) and their properties (melting temperature, thermal conductivity, etc.), the welding intensity, pressure, etc.

The goals to be achieved can be as varied as our imagination. The factors may depend, among other things, on the process chosen to achieve the objective. It is important to know that there are different methodologies for DOEs according to the objective to be reached [KAR 04]; some examples are as follows:

1) I am seeking to identify the influential factors of a new multi-parameter process → I will conduct a *factor screening* DOE.

2) I seek to quantify factors and predict their influence → I will conduct a *modeling* DOE.

3) I seek to obtain the best operating condition for my process → I will conduct an *optimization* DOE.

4) I try to take into account environmental factors, equipment wear, etc. → I will conduct a *robustness* DOE.

In this chapter, the DOE approach is described for the case study on the optimization of the resistance soldering process [BID 16]. *The aim is to model and optimize my process.*

At this stage, it is agreed that the operative and metallurgical weldability are demonstrated. The choice of the joining technique and the materials to be joined are not questioned in the DOE.

The optimization of my process takes place in seven steps, as shown in Figure 3.10.

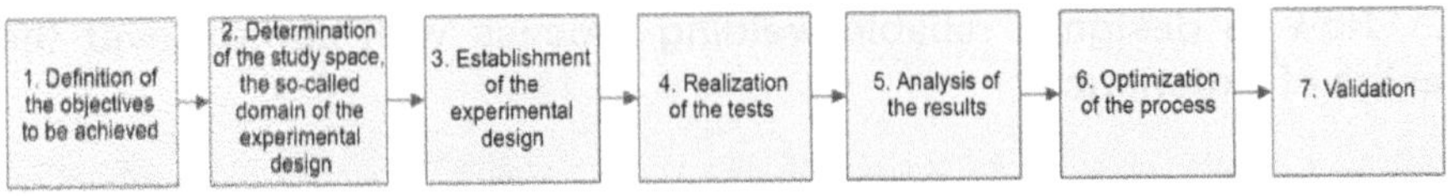

Figure 3.10. *Diagram showing the steps of the optimization process*

This diagram immediately reminds us of the steps of a DMAIC.

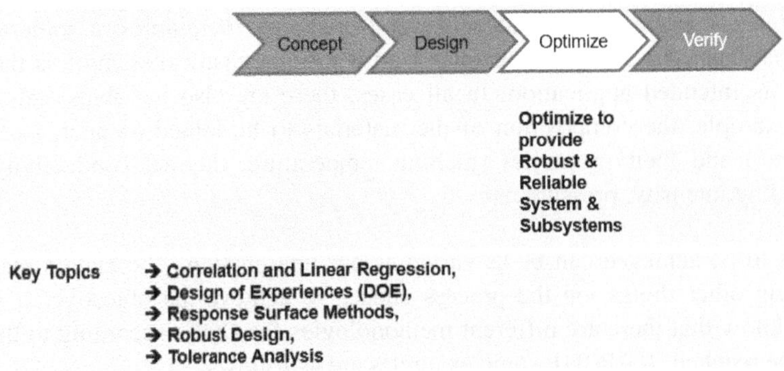

Figure 3.11. *Diagram showing the steps of the DMAIC process*

The DOE therefore follows the DMAIC approach.

3.6. Definition of the objectives

The definition of the objectives to be attained is done by several people, for example, the project manager, the technical manager and the quality department.

It is a question of putting the approach into context by asking several questions detailed below (the answers in italics correspond to the answers provided for the case study).

1) What is the context? *The industrialization of a new interconnector.*

2) Which product(s) are concerned? *The interconnector and the stator.*

3) What is the deadline for the process optimization? *Expected completion before JPC, and for JPC with optimized processes.*

4) Where is the DOE to be carried out? *On the production line → provide slots on the line for carrying out the tests.*

5) Why carry out a DOE? *To guarantee the good behavior of the soldering between the interconnector and the stator during the life of the product (customer request).*

3.6.1. *Determining the study space*

3.6.1.1. *Factors*

The first step is to make an exhaustive list of the factors that can influence the quality of the process. It is recommended to classify them according to the Ishikawa diagram (see Figure 3.12).

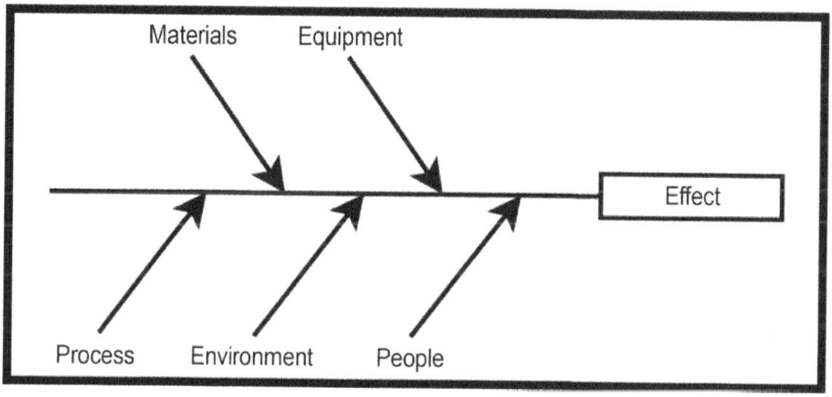

Figure 3.12. *Ishikawa diagram (source: Wikipedia)*

Next, we will determine which factors will become experimental factors by classifying them according to whether they are:

– studied: the DOE variables;

– controlled: the fixed variables of the process parameters;

– uncontrolled: the variables that can change from one trial to another that are not controlled;

– blocks: variables that can be sorted as differences from one day to another.

Table 3.2 shows the approach for the case of optimizing the resistance soldering process in the example.

Factor type	Description	Factor classification
Material	Welding equipment used: Generator and transformer, shunts, electrode holder	Controlled
	Cooling system	Controlled
	Electrode geometry	Controlled
Material	Interconnector material	Controlled
	Stator material	Controlled
	Soldering filler material	Controlled
	Material of the electrodes	Controlled
	State of wear-and-tear (number of soldering operations already performed)	Controlled
Medium	Temperature	Uncontrolled
	Hygrometry	Uncontrolled
	Cleanliness of the rooms	Uncontrolled
Method	Intensity	Studied
	Pressure	Studied
	Time	Studied
	Slope of rise in intensity	Controlled
	Electrode closing time	Controlled
	Electrode holding time	Controlled
	Relative position of the interconnector in relation to the stator	Uncontrolled
Workmanship	Manual electrode mounting	Block

Table 3.2. *Factor classification*

3.6.1.2. *Field of study*

The factors studied thus determine the space of the DOE (Figure 3.13), but there are still elements missing in order to obtain the study domain (Figure 3.14).

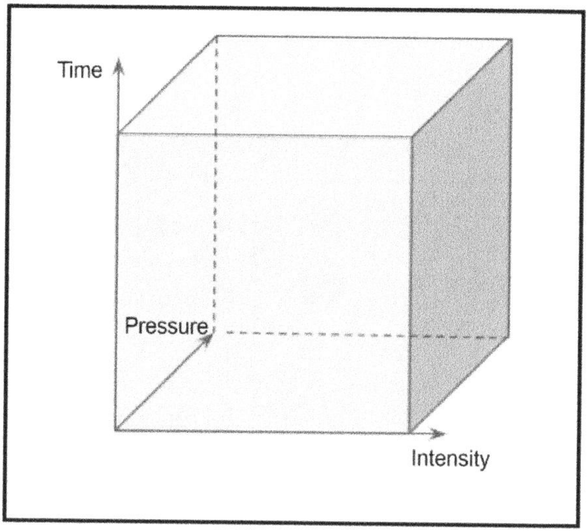

Figure 3.13. *Design of experience space*

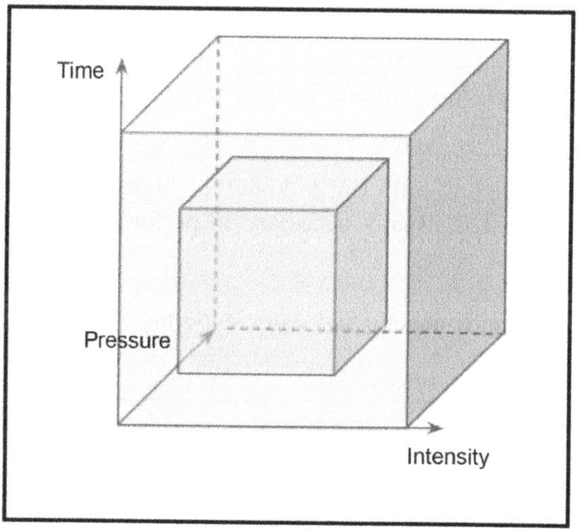

Figure 3.14. *Design of experience area (in blue)*

To determine the domain of the study, the so-called high and low configurations of the DOE must be known (Figure 3.15).

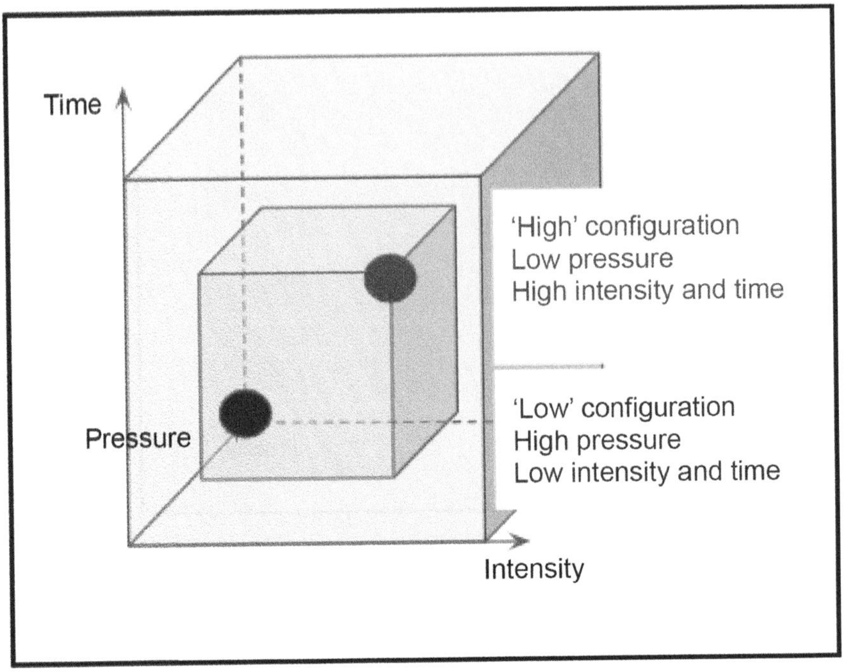

Figure 3.15. *Design of experience domain (in blue) and high and low configurations*

Sometimes experience is sufficient to know the high and low configurations of the DOE domain, but it is usually necessary to perform some tests to determine these limits.

In the case of the interconnector and stator soldering, the configurations obtained are detailed in Table 3.3.

Factors	Low configuration	High configuration
Time [ms]	100	600
Intensity [kA]	4	7.5
Pressure [bar]	5	3

Table 3.3. *Details of the high and low configurations for the design of experience*

3.6.1.3. *Answers to optimize*

At this stage, the answers have not yet been described, but they will determine the outcome of the optimization process.

In the process, we must ask ourselves the question: What are we trying to achieve?

In the case of the example, the mechanical strength of the soldered assembly is an answer, but it is not the only one. In addition, there are visual criteria related to the indentation of the electrode, the wetting of the filler metal or the non-degradation of the plastic casing, for example.

For each answer, a success criterion must be determined. Table 3.4 shows the optimization criteria for the interconnector soldering.

Description	Criteria	Optimization criteria
Mechanical resistance (destructive)	Resistance > 400 N	Maximum
Visual aspect after welding (non-destructive)	1) No visible indentation 2) Slight indentation 3) Pronounced indentation	2) Slight indentation
Visual aspect of the fracture surface	1) Fracture in the soldered area (interconnector/stator separation) 2) Fracture in the middle of the soldered area 3) Fracture outside the soldered area	3) Fracture outside the soldered area

Table 3.4. *Optimization criteria for solder strength*

3.6.1.4. *Choosing the DOE model*

In the context of the DOE optimization, the choice of a test matrix based on the analysis of response surfaces is appropriate.

The name of response surface denotes the concern to visualize, through a geometrical representation, what the response of a physical process to stimuli is [DUP 12]. The response studied Y results from the application of a transfer function that characterizes the system with the input parameters (Figure 3.16).

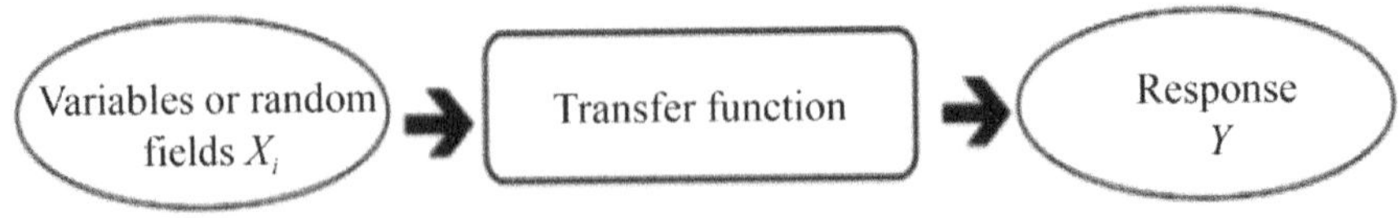

Figure 3.16. *Response of a transfer function to a stimulus modeled by random variables or fields [DUP 12]*

Polynomial response surfaces are mostly used in mechanics [DUP 12]. The polynomial quadratic model is a simple polynomial that describes the relationship between the factors studied and the responses. For a response Y (e.g. mechanical strength), the relationship with the factors (e.g. time A and intensity B) can be written according to equation [3.1].

$$Y = \beta_0 + \beta_1 A + \beta_2 B + \beta_3 AB + \beta_4 A^2 + \beta_5 B^2 \tag{3.1}$$

where the response surface coefficients β_x are determined by the least squares method from the DOE. An illustration of the response surfaces is shown in Figure 3.17.

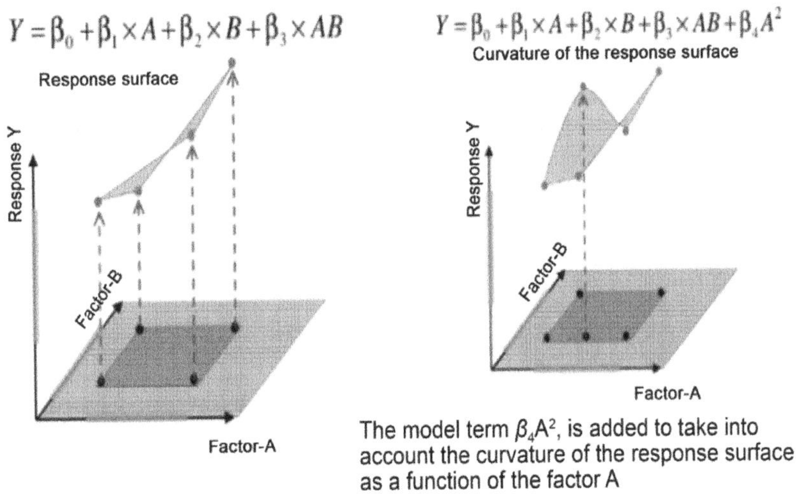

Figure 3.17. *Illustration of the response surfaces [VIV 16]*

This type of model is adapted for the case where the domain is well defined and allows us to reduce the number of tests to be carried out.

3.6.2. *Building the DOE*

The use of Box–Behnken designs has the advantage of reducing the number of test points compared to central composite designs (CCD) [MIN 18]. Other constructions are possible; as described in [FRA 08], the objective always being to obtain a reliable response at a lower cost. For each factor, the Box–Behnken design takes into account only three levels of the DOE domain defined upstream. It never includes points at the extremes of each factor. These matrices are then completed with a point at the center of the domain. An illustration of the points to be made is presented in Figure 3.18. The black points are the test points to be carried out in the study domain, itself represented in blue and determined in section 3.2.

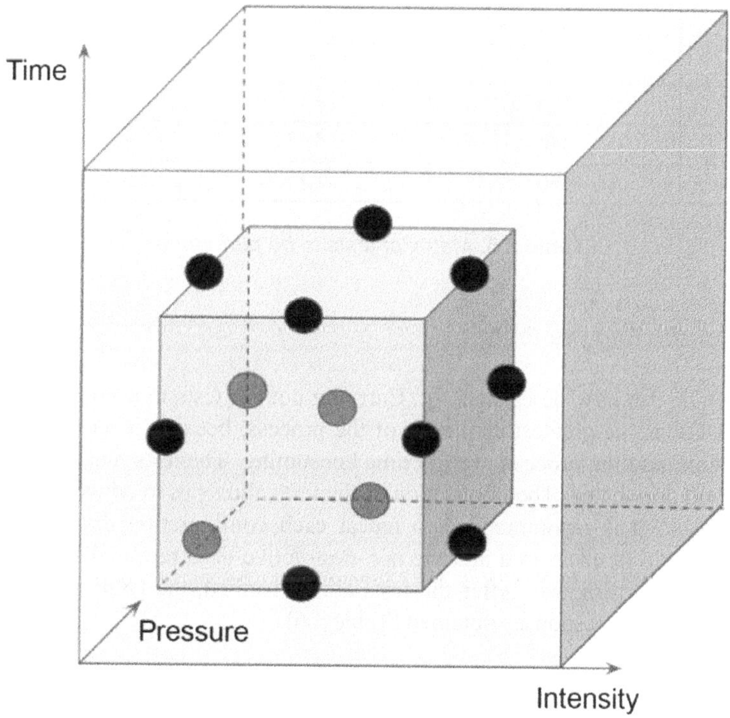

Figure 3.18. *Box–Behnken design, illustration of the points to be achieved*

Software such as Design Expert or Minitab can help determine the DOE matrix.

For the purpose of this example, the test matrix is detailed in Table 3.5.

Run	Factor 1 A: Time [ms]	Factor 2 B: Intensity [kA]	Factor 3 C: Pressure [Bar]
1	600	5.75	5
2	600	7.5	4
3	600	4	4
4	350	7.5	5
5	100	7.5	4
6	350	4	5
7	100	5.75	5
8	100	5.75	3
9	600	5.75	3
10	350	7.5	3
11	100	4	4
12	350	5.75	4
13	350	4	3

Table 3.5. *Matrix of tests to be performed*

3.6.3. *Conducting the tests*

The DOE can now be carried out. Carrying out the tests in accordance with the defined DOE is the greatest challenge of the process, because it requires rigor and organization, and the process itself is time consuming. The tests must be performed in order and duplicated. The duplication of the tests allows us to adjust the reliability of the results. It is recommended to repeat each configuration three times. Each response should be measured and the non-destructive tests repeated to improve the reliability of the response. After the tests are performed, the DOE matrix table is completed with the responses obtained (Table 3.6).

Run	Factor 1 A: Time [ms]	Factor 2 B: Intensity [kA]	Factor 3 C: Pressure [Bar]	Response 1 Mechanical strength [N]	Response 2 Visual before fracture	Response 3 Visual after fracture
1	100	5.75	5	478	1	1
2	100	5.75	3	710	2	3
3	350	7.5	3	912	3	3
4	600	5.75	3	824	2	3
5	600	5.75	5	530	2	2
6	350	5.75	4	798	2	3

7	350	4	3	723	2	2
8	600	7.5	4	800	2	3
9	350	7.5	5	552	2	1
10	100	4	4	451	1	1
11	600	4	4	813	2	3
12	100	7.5	4	623	2	2
13	350	4	5	598	2	2

Table 3.6. *Design of experience complete with responses*

3.6.4. *Analyzing the results*

As described above, the analysis of the results is done using a quadratic model. For each answer:

– visual aspect;

– mechanical resistance;

– aspect of the fracture;

corresponds to a law satisfying the quadratic polynomial model (see equation [3.1]).

The complexity of the model for each response can be reduced according to the coefficients obtained; in the case where the coefficients governing the curvature terms are almost zero, the model is simplified. The use of DesignExpert or Minitab allows us to process the results in order to obtain the behavior laws. The analysis of the model is then based on the analysis of the r_{ij} residuals. *The residual is the difference between the measured response Y_{ij} (for the replication j of test number i) and the value of the response predicted by the model Y_i.*

$$r_{ij} = Y_{ij} - Y_i \quad [3.2]$$

The objectives for each answer are as follows [VIV 16]:

– The normal plot of the residuals shows that the residuals follow a normal distribution that describes the random variation of the process.

– The residuals with respect to the predicted values show random dispersion, meaning that the same level of accuracy is expected for small and large values of the response.

– The residuals versus run order are also random, meaning that the study environment did not change during testing.

– The predicted versus actual value is distributed along the 45° axis and indicates a better prediction for a higher mechanical strength.

In this case, the models are accepted and the optimization process can be started.

3.6.5. *Process optimization*

The optimization of the process is obtained using a desirability function. For each answer Y_k, a desirability function $d_k(Y_k)$ assigns a value between 0 and 1 as the possible value of the answer: 0 is the least desirable answer, and 1 is the most desirable answer that meets all the process optimization criteria. In the example, the objective of the experimental design is to have:

– maximized mechanical strength of more than 400 N;

– visual aspect after welding at 2 (slight indentation visible);

– visual aspect of the fracture surface at 3 (fracture outside the soldered area).

The individual desirabilities are then combined using a geometric average:

$$D = \prod_k \quad d_k(Y_k)^{W_k}$$

where W_k is the relative importance given to the indexed criterion k, such that:

$$\sum_k \quad W_k = 1$$

D = 0 if one of the d_k is equal to zero and D = 1 if all $d_k = 1$.

The optimum configuration corresponds to the configuration giving the highest D.

The result obtained in the case of resistance soldering is presented in Table 3.7.

Time [ms]	Intensity [kA]	Pressure [Bar]	Mechanical durability [N]	Visual before break	Visual after break	Desirability
484.622	4.000	3.000	813.922	2.058	3.000	0.913

Table 3.7. *Results obtained for resistance soldering*

3.6.6. *Validation*

The optimum configuration is often a configuration located outside the matrix of the experimental design. Indeed, the optimum configuration (the red point in Figure 3.19) is not a point in the DOE matrix. Therefore, this optimum must be confirmed and validated.

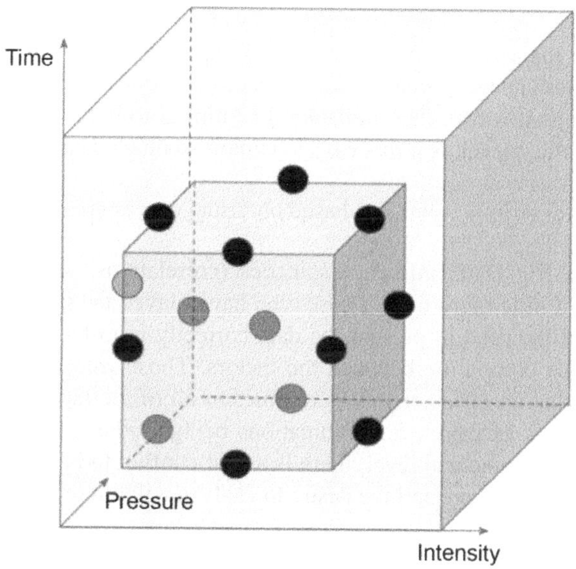

Figure 3.19. *Visualization of the optimum (in red) in the design of experience domain*

The practice, in the case of welding, is to repeat the welds 30 times in the optimum configuration and in the process cycle-times, and to confirm that the mechanical strength is always higher than 400 N.

3.7. Big Data and process?

Collecting large quantities of digital data in central databases results in added value when you know how to exploit them. Numerous jobs are appearing every day in this sector: data scientists, programmers in machine learning algorithms [GER 19], deep learning [GER 20], artificial intelligence with or without neural networks. Many actors are now offering different business models in this booming market, starting with the algorithmic coding of a production line and ending with the predictive modeling of a complete factory via a simple website, with the possibility of constantly questioning which factors influence which output characteristics with a certain expected tolerance. The experience feedback of these specialists, based on Big Data, underlines that *the main source of industrial variability in a general way is explained as 10% from the material, 20% from the machines and 70% from people.*

Usually, the main causes of loss in overall equipment effectiveness (OEE) are non-quality, series changes, organization, breakdowns, micro-stops; thus, it has been shown, in the digital world, that humankind continues to hold destiny in their own hands. On a practical level, is it that easy to exploit so much data?

In Big Data, descriptive statistics based on visual curves are easy to manage.

However, in Big Data, inference statistics (correlations, search for influential factors) require a little more time. Those who have played the interactive game will have understood the need to present the data correctly in order to be able to easily deduce a possible correlation between the factors. Those who have used statistical software know that an efficient way is to position all of the factors in columns and have as many rows as there are combinations being studied, so as to not have to rework them. At the practical level, there is a first conflict to be solved between the practical capacity of storage and the desire to analyze without reprocessing:

- by definition, the number of factors to be studied is never really known before any question of influence is asked, and besides, the number of connected objects or variables to be stored is infinite and in perpetual motion!

- on the contrary, any computer developer must format a table with a number of predefined and thus fixed columns (besides, some software are limited in terms of numbers).

One of the tricks commonly used is the stacking of various factors in the same column, but we see here one of the great difficulties of Big Data: the formatting of the storage requires gymnastics to be able to correctly study the influences between factors, which is a source of waste of energy, time and, globally, resources.

The authors have therefore deliberately chosen a formatting adapted to any correlation study without the waste of a reprocessing behind it.

Numerous software exists to explore a Big Data environment through machine learning algorithms.

For example, SALFORD (MINITAB Group) will propose a predictive model on 80% of the sample and test it on 20% of the residual values.

Here, some images, taken on the fly, have been obtained from an analysis of an MES system at a real-world site.

Among more than 100 variables, we can notably highlight which ones have the most influence on an output variable.

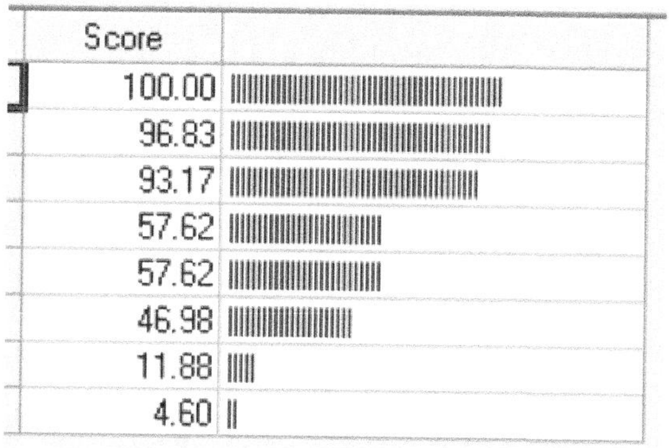

Figure 3.20. *MES image showing influence on output variables*

It is also possible to study the influence curves by visualizing all of the nonlinear effects, which is required when studying complex phenomena over a long period of time.

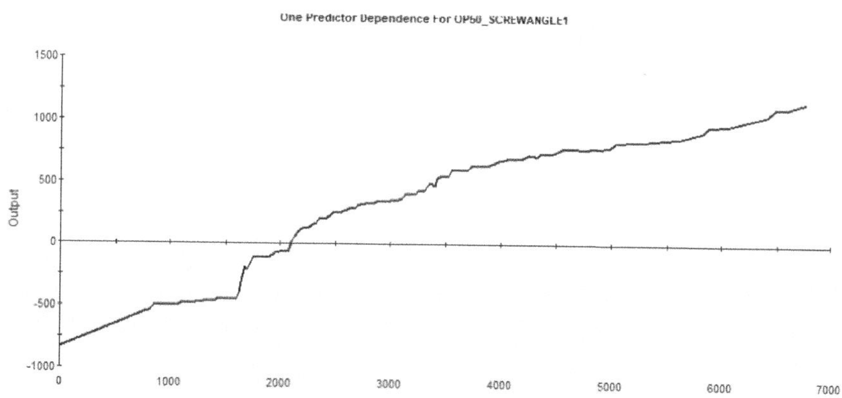

Figure 3.21. *Image showing one predictor over time*

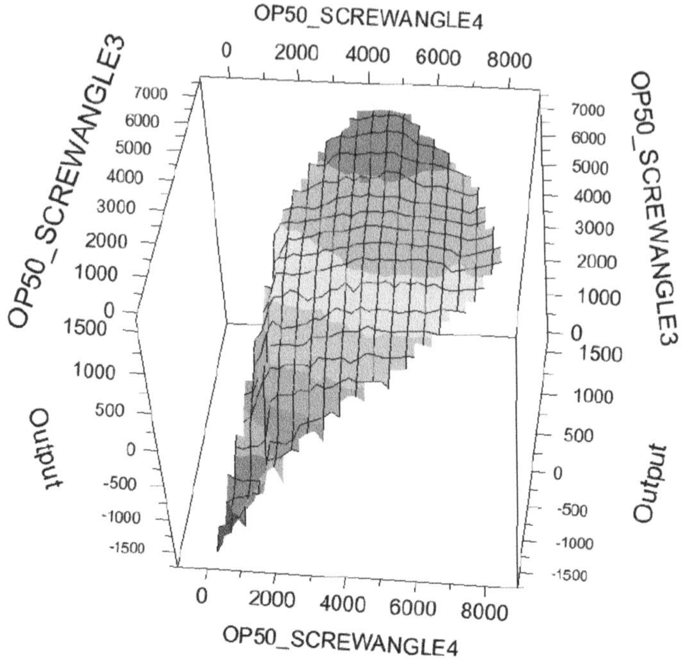

Figure 3.22. *Visualizing correlations through graphs*

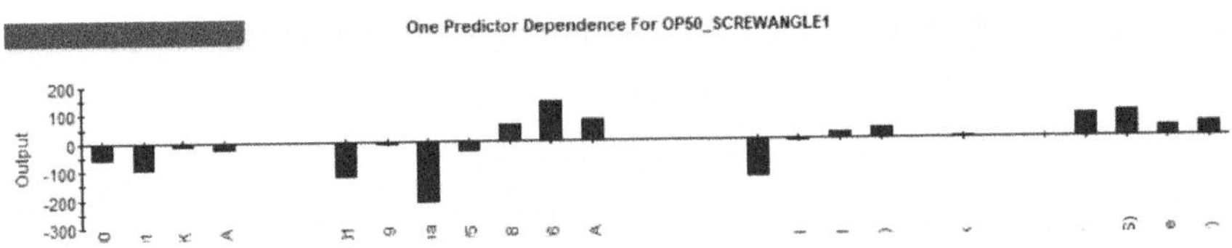

Figure 3.23. *Visualizing the effects of variables on an output: the larger the bar, the greater the effect (positive or negative)*

The idea here is to fully exploit predictive models that will improve the OEE and reduce scraps. In machine learning, there are other possibilities with TENSORFLOW, which is also widely used. For example, here we can see a study done in PYTHON language where some references on the chart have more isolated points than others.

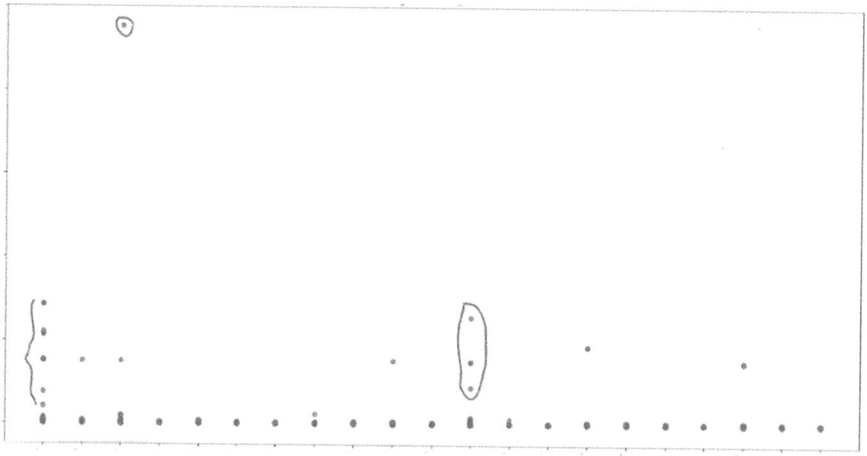

Figure 3.24. *Chart showing isolated reference points*

It is then sufficient to find the event concerned, or better, to merge these data with another database to immediately identify the root cause. Even if machine learning and deep learning require some practice and targeted algorithmic knowledge (see [GER 20] and [BID 16]), the authors specify that classical statistical tools can also capture order ascending data from several years, to always build a statistical model in order to predict the behavior of certain variables, and thus to better optimize them.

In Appendix 3 (see section 3.12), an example of what we can do with traditional statistics is given. With the practice of Big Data exploding, what is the future of Lean Six Sigma?

This is the current theme of the major LEAN institutions (Lean Six Sigma University, LEAN France Institute, etc.).

Big Data can recover everything; it can be associated to a cylindrical pipe. Lean Six Sigma, as practiced in our interactive game, can be compared to a funnel because of its progressive process of factor selection. At first glance, these two notions would be antinomic. Some even predict the death of DMAIC and Lean Six Sigma. However, in this article we have discussed the importance of formatting data beforehand, for immediate analysis afterwards, to avoid any subsequent MUDA reprocessing.

The skills of Lean Six Sigma experts still have good days ahead, to help build and take part in the future revolution of cylindrical pipe data analysis, without time-consuming reprocessing. In Big Data, machine learning practitioners know the difficulty of having diagonal confusion matrices, and must often:

– balance the data classes, at the expense of the initial quantity (hence the importance of knowing how to choose them, especially in stratified sampling, this is Lean Six Sigma!);

– introduce more factors, hence the need to know how to identify them (this is Lean Six Sigma too!).

The added value of Lean Six Sigma experts in a world of data scientists is thus the knowledge of the field, and more specifically the knowledge of which databases from different sources need to be connected in order to identify the root causes.

All this with the shared objective of reducing scraps and increasing the OEE.

3.8. Conclusion

In response to the question of how to set up an efficient methodological system to design innovative, competitive and robust products and processes, the Lean Six Sigma method remains essential, especially in its DFSS variation, which is not the subject of a standard and is still largely unknown and underused.

Efficiency comes from the common language between all of the actors, and from knowing how to exploit data, especially Big Data.

The main attraction of a Lean Six Sigma method is that it easily allows for the integration of external companies that do not yet know the host company.

The robustness comes from the non-variation, as mentioned and studied through our DMAIC case study. This is because variation is still very expensive.

Innovation obviously remains highly dependent on the pioneering spirit of experts and people, who, according to the first returns in Big Data, are ultimately the timeless vector of any source of competitiveness.

3.9. Appendix 1: example of an ANOVA study

Let us take a look at a recent example, where we have interactions between factors that give us a model with an R2 of 100%; in other words, a perfect model:

Regression analysis: VAR1 versus VAR2; ... i8; VAR8r

Method

Lines not used	11

Normalization of continuous predictors

Subtract the mean and divide by the standard deviation

Predictor	Average	SD
VAR2	54.0313	2.73042
VAR3	54.9317	2.17448
VAR4	56.6300	2.75049
VAR5	54.5769	3.34880
VAR6	55.7354	1.92814
VAR6r	54.9417	2.63201
VAR7	55.4088	1.98267
VAR8	54.1324	2.98271
VAR8r	54.7675	2.47161

Step-by-step term selection

α to include = 0.15; α to exclude = 0.15

The step-by-step procedure added terms to maintain a hierarchical model at each step.

Analysis of variance

Source	DL	SomCar adjust	CM adjust	F value	p-Value
Regression	27	146.216	5.41539	1.36306E+11	0.000
VAR2	1	2.290	2.29031	5.76472E+10	0.000
VAR3	1	0.037	0.03691	9.29109E+08	0.000
VAR4	1	0.758	0.75840	1.90889E+10	0.000

VAR5	1	0.052	0.05161	1.29909E+09	0.000
VAR6	1	0.028	0.02830	7.12394E+08	0.000
VAR6r	1	0.530	0.52997	1.33393E+10	0.000
VAR7	1	1.540	1.53983	3.87576E+10	0.000
VAR8	1	0.187	0.18748	4.71893E+09	0.000
VAR8r	1	0.007	0.00748	1.88305E+08	0.000
VAR6r*VAR6r	1	3.405	3.40548	8.57162E+10	0.000
VAR7*VAR7	1	1.612	1.61155	4.05628E+10	0.000
VAR2*VAR4	1	3.299	3.29916	8.30401E+10	0.000
VAR2*VAR5	1	0.475	0.47509	1.19580E+10	0.000
VAR2*VAR6	1	0.003	0.00348	87539723.96	0.000
VAR2*VAR7	1	0.102	0.10247	2.57917E+09	0.000
VAR2*VAR8	1	0.379	0.37864	9.53032E+09	0.000
VAR2*VAR8r	1	0.873	0.87255	2.19621E+10	0.000
VAR3*VAR5	1	0.000	0.00004	1013133.81	0.001
VAR3*VAR6r	1	2.838	2.83810	7.14352E+10	0.000
VAR3*VAR8	1	1.635	1.63530	4.11605E+10	0.000
VAR4*VAR5	1	2.213	2.21337	5.57107E+10	0.000
VAR4*VAR8	1	0.050	0.04968	1.25054E+09	0.000
VAR4*VAR8r	1	2.300	2.30011	5.78938E+10	0.000
VAR5*VAR6r	1	0.000	0.00031	7690783.63	0.000
VAR5*VAR7	1	0.028	0.02799	7.04542E+08	0.000
VAR6r*VAR8r	1	0.000	0.00000	68218.53	0.002
VAR8*VAR8r	1	0.030	0.02964	7.46003E+08	0.000
Error	1	0.000	0.00000		
Total	28	146.216			

Model summary

S	R squared	R squared (adjust)	R squared (prev)
0.0000063	100.00%	100.00%	100.00%

Coded coefficients

Term	Coeff	Coef ErT	T value	p Value	FIV
Constant	56.9936	0.0000	8,937,986.52	0.000	
VAR2	–0.772372	0.000003	–240,098.38	0.000	7.29
VAR3	–0.142044	0.000005	–30,481.30	0.000	15.30
VAR4	0.649829	0.000005	138,162.66	0.000	15.59
VAR5	0.182550	0.000005	36,042.93	0.000	18.08
VAR6	0.093627	0.000004	26,690.71	0.000	8.67
VAR6r	0.468710	0.000004	115,495.94	0.000	11.61
VAR7	0.747254	0.000004	196,869.37	0.000	10.15
VAR8	–0.259533	0.000004	–68,694.44	0.000	10.06
VAR8r	–0.062491	0.000005	–13,722.42	0.000	14.62
VAR6r*VAR6r	–0.612035	0.000002	–292,773.25	0.000	7.21
VAR7*VAR7	–0.698123	0.000003	–201,402.11	0.000	15.79
VAR2*VAR4	1.78056	0.00001	288,166.79	0.000	31.31
VAR2*VAR5	0.488923	0.000004	109,352.56	0.000	11.08
VAR2*VAR6	–0.042023	0.000004	–9,356.27	0.000	14.66
VAR2*VAR7	–0.254899	0.000005	–50,785.49	0.000	13.87
VAR2*VAR8	0.376577	0.000004	97,623.34	0.000	13.65
VAR2*VAR8r	–0.756225	0.000005	–148,196.06	0.000	20.77

VAR3*VAR5	0.005341	0.000005	1,006.55	0.001	19.38
VAR3*VAR6r	0.963642	0.000004	267,273.58	0.000	8.50
VAR3*VAR8	1.44135	0.00001	202,880.53	0.000	30.24
VAR4*VAR5	–0.922253	0.000004	–23,6031.20	0.000	9.52
VAR4*VAR8	–0.193123	0.000005	–35,362.96	0.000	9.31
VAR4*VAR8r	–1.31842	0.00001	–240,611.28	0.000	16.95
VAR5*VAR6r	0.015474	0.000006	2,773.23	0.000	15.17
VAR5*VAR7	–0.121079	0.000005	–26,543.21	0.000	15.21
VAR6r*VAR8r	0.000951	0.000004	261.19	0.002	8.38
VAR8*VAR8r	–0.214743	0.000008	–27,313.05	0.000	36.66

Regression equation in uncoded units

VAR1 = –205.672 – 9.93982 VAR2 – 21.3859 VAR3 + 4.78632 VAR4 + 3.70925 VAR5

+ 0.479845 VAR6 + 0.533327 VAR6r + 23.5970 VAR7 – 11.8644 VAR8

+ 18.5809 VAR8r – 0.088349 VAR6r*VAR6r – 0.177596 VAR7*VAR7

+ 0.237093 VAR2*VAR4 + 0.053471 VAR2*VAR5 – 0.007982 VAR2*VAR6

– 0.047086 VAR2*VAR7 + 0.046240 VAR2*VAR8 – 0.112058 VAR2*VAR8r

+ 0.000734 VAR3*VAR5 + 0.168374 VAR3*VAR6r + 0.222230 VAR3*VAR8

– 0.100127 VAR4*VAR5 – 0.023540 VAR4*VAR8 – 0.193939 VAR4*VAR8r

+ 0.001756 VAR5*VAR6r – 0.018236 VAR5*VAR7 + 0.000146 VAR6r*VAR8r

– 0.029129 VAR8*VAR8r

3.10. Appendix 2: studying the variability of cycle times

For example, here we see that over one year of operating an MES system, a few extreme points, few in number and therefore treatable in the form of QRQC, explain a high average with a strong potential for improvement to keep only a sketch packed at the target level.

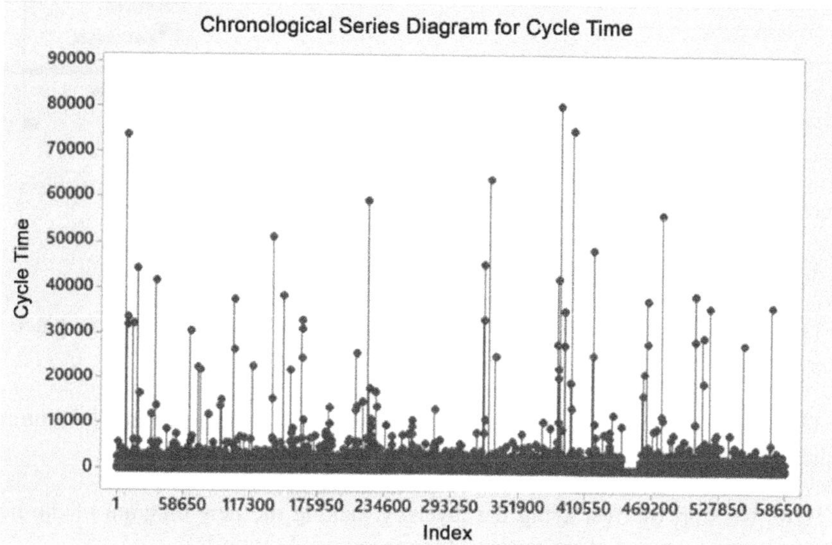

The reliability of a process depends above all on a flow that flows within fixed limits over a long period of time. Two tools are very useful to fluidify operations:

– in organization with rotating operators, the reference tool is the *man/machine diagram*;

– in fixed operator organization, the reference tool is the YAMAZUMI.

The authors propose here to share a practice for a very effective way of using the man/machine diagram:

Starting from the first station, equipped with auto-ejection, the operator will turn from station to station by performing a value addition (VA) operation (in blue), loading it onto a machine, press the start button (in red), take the part already processed, go to the other station by performing a new VA operation during its displacement by recovering another component in supply (in green), and repeat this cycle until the unloading of the customer packaging. We will take care not to have a cycle of less than 50 s to avoid the whirlwind, and less than 4 km per day and per person in daily displacement. Let us see apply a practical case study.

After timing on the ground:

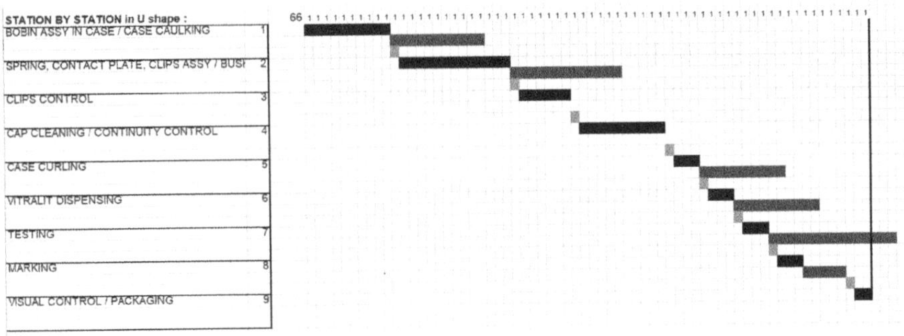

where one cell = 1 s.

A complete loop represents 66 s for one operator.

The question is: What is the maximum number of operators we can place in order to boost the cycle time without the MUDA waiting?

The answer is simple: make a layer of the new graph and superimpose it on top of the first one, and then look at where we have no more collisions.

Here, we start by overlaying the layer by making the new diagram of the next operator start at the end of the cycle of the machine at Station 1:

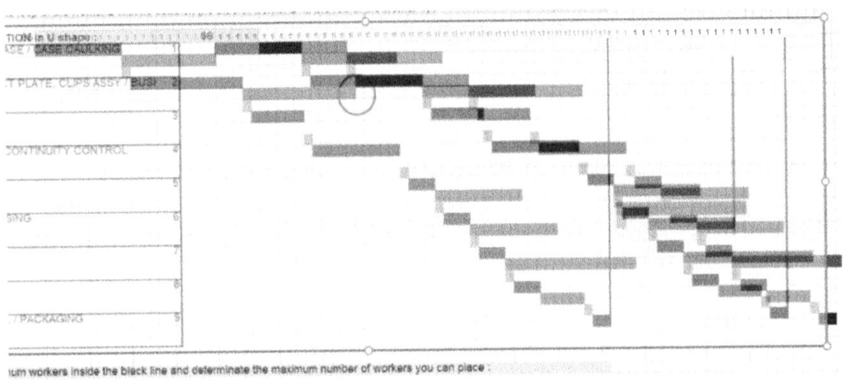

The diagram shows a conflict on item 2. Therefore, we need to superimpose the layer again in order to have no more collisions:

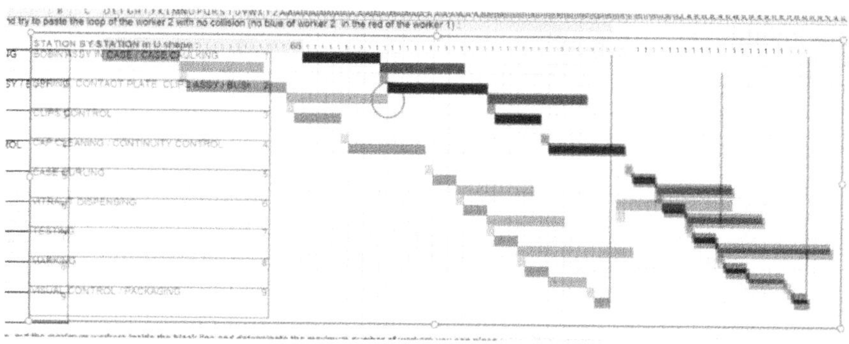

which allows us to deduce the overlap between two operators:

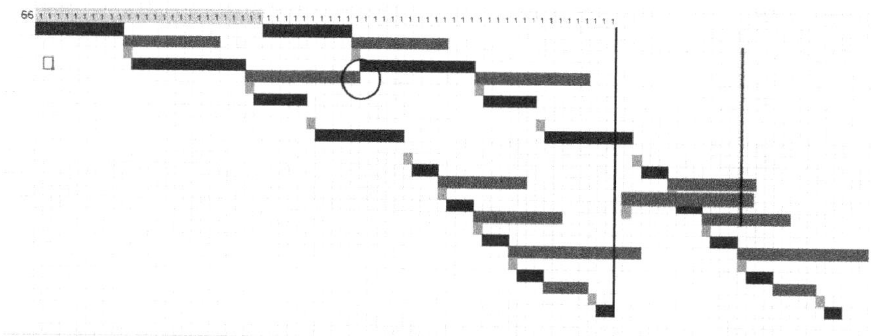

and then reproduce it:

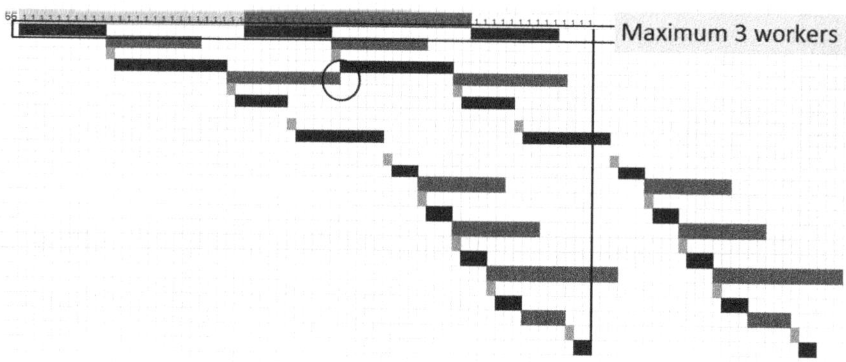

Finally, we can deduce that three operators will be able to produce (without collision/without MUDA WAITING) on this line with a possible variability of the cycle time, which in turn will depend on the number of operators that we will assign:

		TCY
Number of worker =	1	66
Number of worker =	2	33
Number of worker =	3	22

Now a question that may arise is: How do we boost the cycle time for a reduced investment?

Just go through the sequence again.

Nine times out of 10, you will be told that you have to boost the cycle time of Station 1, or even of all the stations. This is false! The collision is taking place at Station 2, and as such, you only need to invest some money in this station to compact the diagram and reduce the complete loop! Next, look again at where the collision will take place and repeat, until you reach the target cycle time (e.g. here we could work on an optimization at 15 s).

As you can see, the optimization here is based on a simple visual layer where we work on superimposed curve conflicts! *In a rotating organization, there is no point in balancing the positions*, contrary to a very widespread idea among Methodists. On the other hand, in a fixed organization, it is important to balance the positions; the second tool concerned is the YAMAZUMI. A concrete example: each task is cumulated over the others (Yamazumi means "to cumulate" in Japanese), that is, to add them station by station in the form of a visual graph:

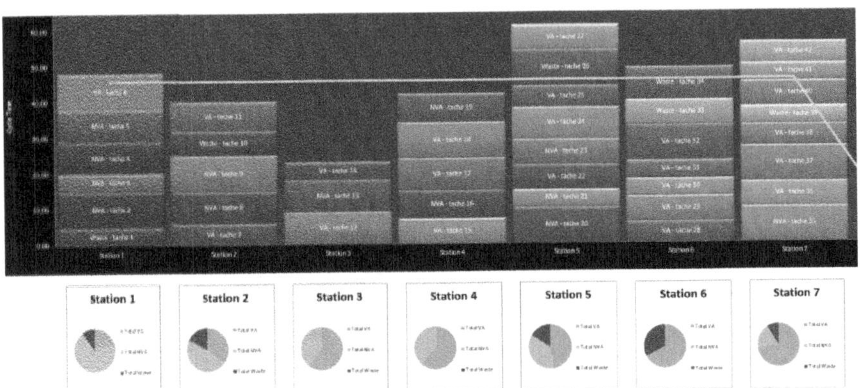

Here, we can clearly see that there is a balancing problem on the line, which has seven stations. This graph also allows us to visualize that there are:

- MUDA (100% wastage);

- value additions;

- no value additions (a choice made during the design phase in order to keep them so as to eliminate more MUDAS, but which does not bring added value, in a word, they are so-called compromise MUDAS that are left to eliminate more elsewhere, in greater proportions).

The next step will be to kill the MUDAS (which will shift the tasks and give us a new YAMAZUMI). Next, we will have to rethink the sequence, that is, to shift some tasks with high added value upstream or downstream to balance all of the stations! Therefore, we can anticipate (if possible) a task from Station 5 to Station 3 in order to rebalance the cycle time and suggest another sequence with more modifications for a better final balancing. For example, if we work in this way, we can end up with:

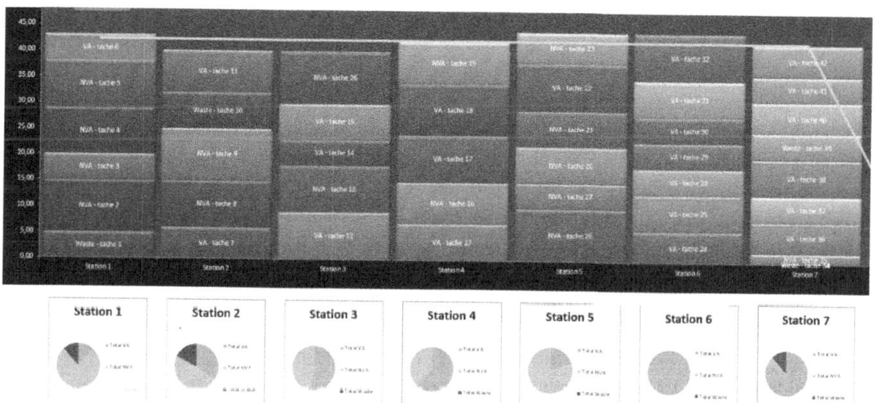

So why include this type of tool in an article about reliability?

Simply because the first source of variation in a poorly designed process will be the cycle time. Here is a real-life example of a line where the first operator station is loading products onto a pallet, some of which are front loaded and some of which arrive in bins and are mounted at the correct ergonomic height by a *Bloksma*. A recent study showed that the first source of variability came from the frequency of bin changes in the *Bloksma*, which in turn had a severe impact on the performance of the line (through a ricochet effect):

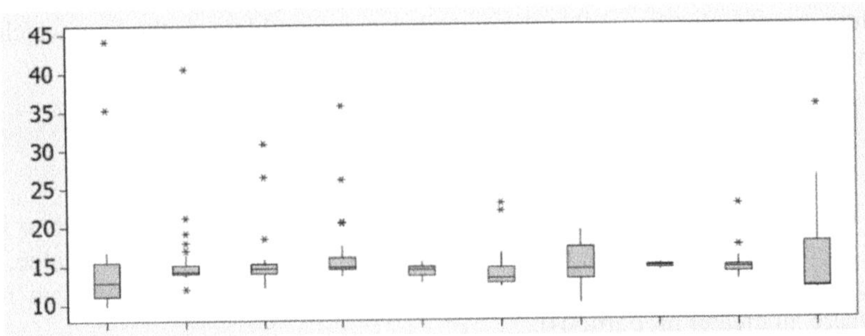

On the box plot, we see abnormal points that occur during bin changes in a frequent manner, which immediately propagates to stations downstream. We also see this on the confidence interval diagram:

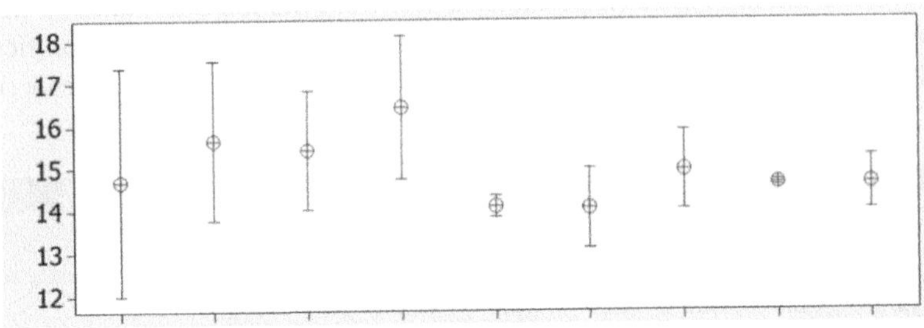

Conclusion: If the impact of a bin change is not correctly integrated into the study from the outset, the line will never achieve the expected results because positioning a front-end supply afterward may simply be impossible if all of the items have been compacted into an allocated area. We also remind you that the four major sources of flow stoppage are as follows:

– non-quality (hence the importance of controlling any source of variability);

– transport (which does not transform, although we have already seen that it is possible to move an operator while adding value in the context of a rotating organization);

– stocks (hence the importance of "one piece flow", i.e. working piece by piece) to be at the right level of the demand, and, once again, to eliminate any form of variability in the process (in order to avoid dispersions due to the use of different tools);

– waiting times (hence the importance of upstream YAMAZUMI and/or man/machine diagrams). We can see several successive months of production of a plastic part, which may come from 4 A/B/C or D prints, that there are discrepancies. On the right, we also see that a change in material can reduce dispersion.

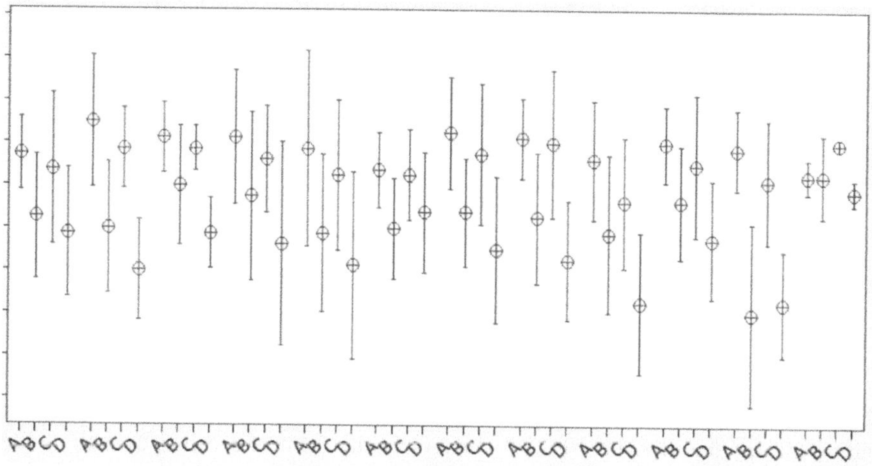

In order to do preventive work, digital simulation tools are very useful. Tools such as PROCESS SIMULATE from SIEMENS or DELMIA from DASSAULT allow us to start from the product and deduce an assembly sequence, then build a flow (i.e. we add how to supply the raw material onto the line), and define the operations, very precisely, with tools (already presented above) for balancing, by working on GANTT. It is possible to simulate a robotic or human sequence and study ergonomics in predictive mode:

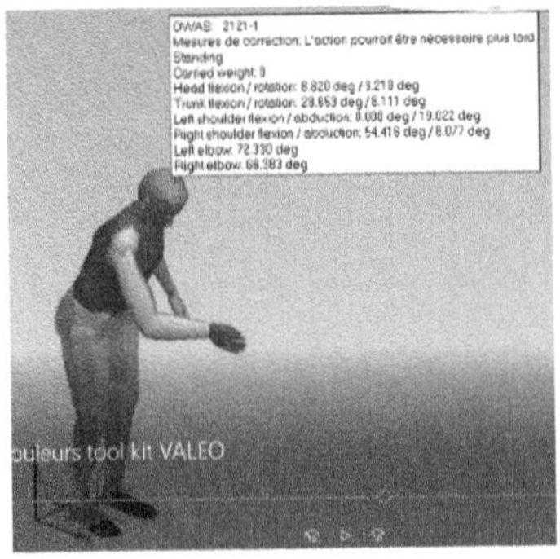

However, experience shows that nothing beats one scale for very effective ergonomic and flow studies. This is why virtual reality, in the form of suitcases that can be transported anywhere in the world or in the form of a cellar, as well as the use of holograms, is now used for the design of new processes.

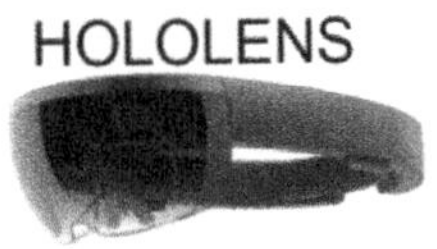

The author's experience has shown that for an ideal use of the MICROSOFT HOLOLENS glasses, the ideal number of triangles (facets) of the 3D file of the machine to be studied must be at most 150,000 triangles, based on a curve from an experiment; see the graph below:

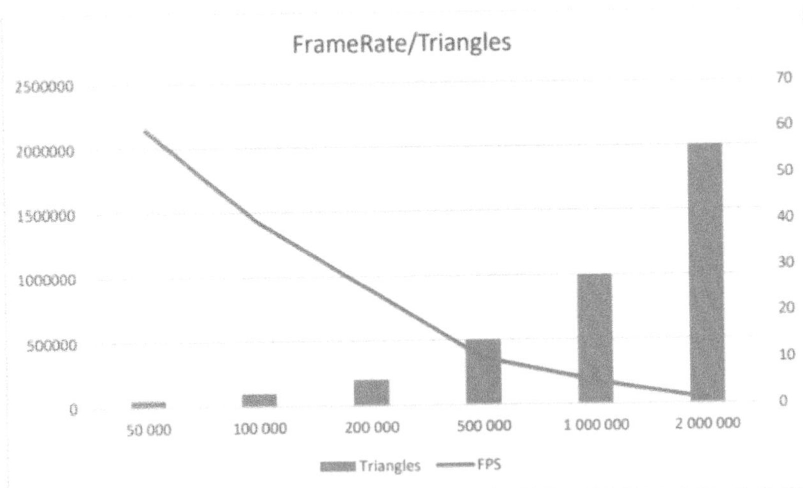

The vital limit of use is 30 fps for a Hololens experience that does not result in flashes, unpleasant latency effects, unfixed virtual objects, etc. These tools allow us to better work on the optimization of flows in order to eliminate MUDAs in preventive work, make ergonomic studies, launch more precise studies of cycle times (especially in robotics) and solicit the opinions of operators. Fundamentally, these tools allow us to work with all the countries in the world, especially low-cost countries, because "3D" is a marvelous universal language (far more effective than the English language) and constitutes as an absolute test of truth before launching the machine in "reality". Doing preventive work is therefore an essential step in lean. We will see later that it is also important to do predictive work, especially in Big Data.

3.11. Appendix 3: example for the use of traditional statistics in Big Data

Recently, the authors were able to produce a complete analysis of the data in a few hours, which made it possible to detect certain anomalies that would be difficult to intercept with other classic measurement tools (and for good reason: most of these anomalies were seen at night, when the support and management teams are less present). We are talking about 12 million lines of data, representing 6 GB. With a simple traditional statistical analysis tool like MINITAB, we could see a disparity in cycle times depending on the version:

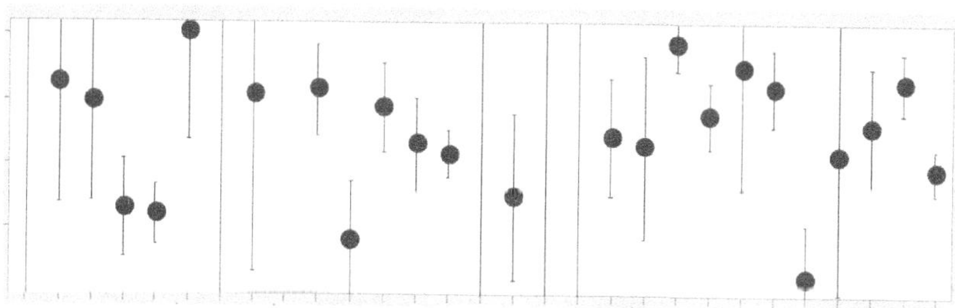

Locate anomalous points:

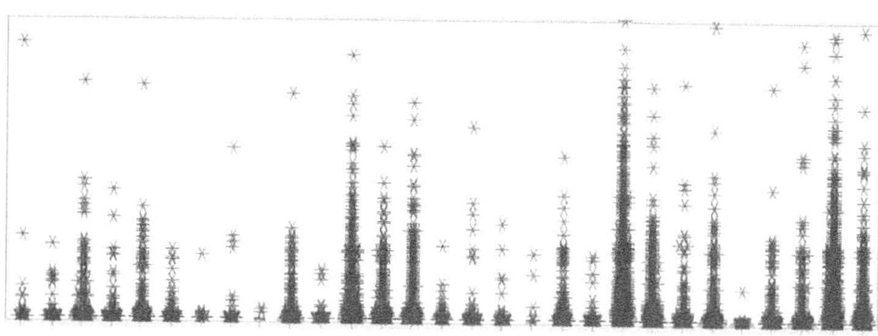

Visualization over the long-term for certain variables:

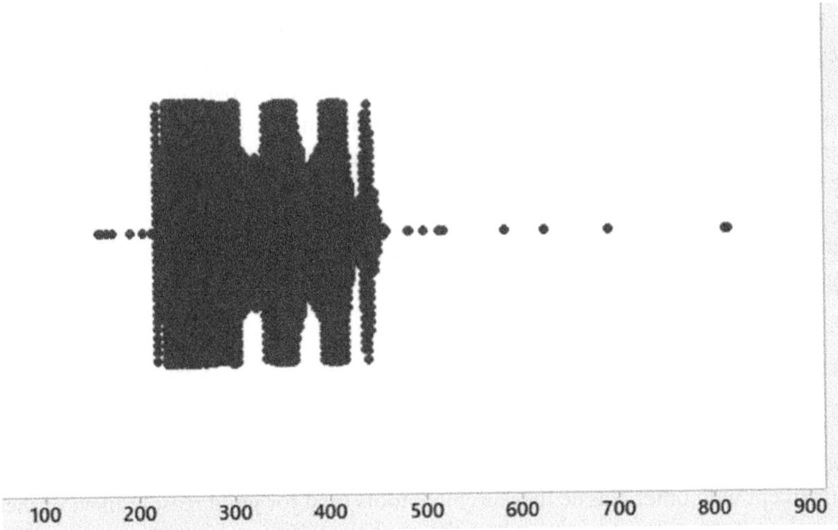

Comparing variables:

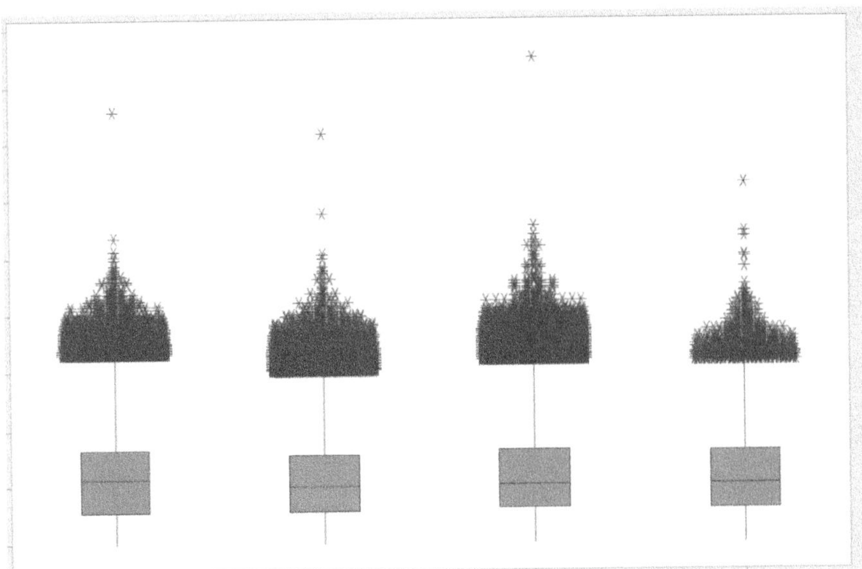

Comparing them by reference:

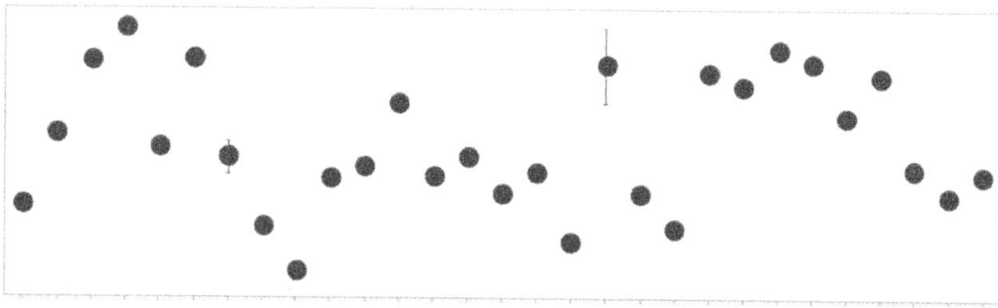

Find correlations between factors with a predictive model with an R2 greater than 87.25% (we will not put the formula here for confidentiality reasons):

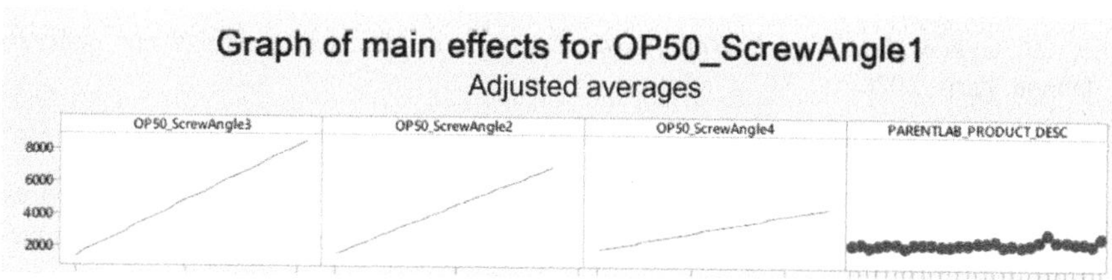

Display surface diagrams between several variables:

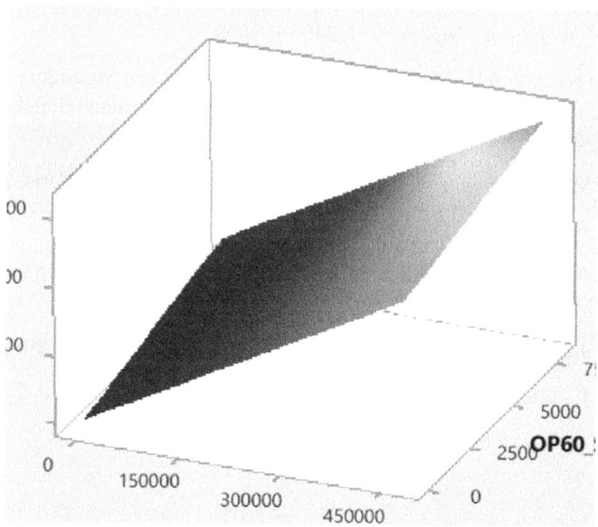

The traditional LEAN 6 SIGMA tools can therefore still be used to explore Big Data and remain complementary in order to draw conclusions that will lead to improved quality and performance.

3.12. References

[AOU 12] AOUDIA H., TESTA Q., *Perfect QRQC*, Maxima, Paris, 2012.

[BAL 14] BALLÉ F., BALLÉ M., *Le Gold-Mine, un récit lean*, L'Harmattan, Paris, 2014.

[BID 16] BIDI L., MASSON P., CICALA E., "Etude de l'influence des paramètres opératoires sur la morphologie des cordons de soudure – cas de soudage hybride (laser/mig)", *ICEMAEP2016*, 2016.

[DUP 12] DUPRAT F., SCHOEFS F., SUDRET B., "Surfaces de réponse physiques et polynomiales", *Fiabilité des ouvrages Sûretés, variabilité, maintenance, sécurité*, pp. 165–193, 2012.

[GER 19] GÉRON A., *Machine Learning avec Scikit-Learn, Mise en oeuvre et cas concrets*, Dunod, Paris, 2019.

[GER 20] GERON A., *Deep Learning avec Tensorflow, Mise en oeuvre et cas concrets*, Dunod, Paris, 2020.

[INS XX] INSTITUT LEAN FRANCE, available at: https://www.institut-lean-france.fr/.

[KAR 04] KARAM S., Application de la méthodologie des plans d'expériences et de l'analyse de données à l'optimisation des processus de dépôt, PhD Thesis, University of Limoges, 2004.

[LEA XX] LEAN SIX SIGMA, available at: https://www.universite-lean6sigma.com/.

[MAT 00] MATHIEU D., PHAN-TAN-LUU R., "Planification d'expérience en formulation : criblage", *Techniques de l'ingénieur*, J2240 v1, 2000.

[NOR 11] NORM NF X06-091, Démarches Lean, Six Sigma, Lean Six Sigma - Exigences des compétences des chefs de projets d'amélioration et des animateurs d'ateliers, Publication number 2-01/04/2011, 2011.

[PIL 13] PILLET M., *Six Sigma, comment l'appliquer*, 2nd ed., Eyrolles, Paris, 2013.

[PIL 17] PILLET M., PILLET D., *Design For Six Sigma*, Eyrolles, Paris, 2017.

[SAB 06] SABRE R., "Planification expérimentale en agroalimentaire", *Techniques de l'ingénieur*, F1005, 2006.

[SMI 14] SMITH J.L., "Phil Crosby's Legacy", *Quality Magazine*, August, 2014

[VIV 16] VIVET L., GSI-RD-H01-0000-424 Process Optimization Standard, 2016.

[WIK XX] WIKILEAN, available at: https://wikilean.com/.

4

Reliability Study of the High Electron Mobility Transistor (HEMT)

This chapter puts forward a methodology to study the reliability of mechatronic systems through the reliability of high electron mobility transistors (HEMTs). We start with an overview of the HEMT technology, its advantages and areas of use. Thermal modeling using the finite element method (FEM) allows for the observation of the thermal behavior of the transistor, and the influence of operating conditions, such as power dissipation and reference temperature, on the operating temperature. The implementation of the FEM is done by the software Comsol Multiphysics. Thereafter, a thermo-reliability coupling is carried out by an integration of the two models: thermal and statistical; the thermal model is developed under Comsol Multiphysics and the reliability model under Matlab. This coupling enables us to estimate the thermal reliability of the HEMT and to identify the influence of the operating conditions on the latter and its performance.

4.1. Introduction

The appearance of mechatronic systems gives rise to phenomena of failure and degradation that develop over time and are not well controlled; it is therefore necessary to detect these failures and to repair them in order to ensure a robust and reliable systems design [ELH 15a].

Among the very important components in these systems is the transistor. It is an electronic element that has created a revolution in the design and development of such systems.

For a color version of all the figures in this chapter, see http://www.iste.co.uk/elhami/applied3.zip.

Chapter written by Abdelhamid AMAR, Bouchaïb RADI and Abdelkhalak EL HAMI.

When we talk about the most developed systems such as airborne systems, we see that most of their failures come from their high-power amplifier (HPA); according to other works, a large percentage of these failures are caused by HEMTs [AMA 19]. These failures are related to the operating temperature, which exceeds the critical values due to the self-heating of this transistor; thus, the operating temperature has a great influence on the performance of the transistor. In order to study this influence, a thermal model based on the FEM is developed under Comsol Multiphysics software to identify the parameters that have the greatest impact on transistor temperature. On the other hand, the first-order reliability method (FORM) and second-order reliability method (SORM) are used to estimate the reliability index and the failure probability. A coupling between Comsol Multiphysics and Matlab has been developed and enables us to link the two models: thermal and reliability.

4.2. HEMT technology

The HEMT based on aluminum gallium nitride/gallium nitride (AlGaN/GaN) is a very important electronic component due to its structure and materials. GaN is characterized by high mobility, a very high electrical breakdown field and high thermal conductivity.

The HEMT appears as a major evolution of the MESFET. However, the difference is that the HEMT uses a heterojunction, that is, a junction between materials with different energy bands, so that the electrons constituting the drain-source current pass through a non-doped semiconductor in order to reduce the transit time and thus increase frequency performance. The electrons' mobility is in fact greater than the lower doping of the semiconductor, because the dispersion of ionized impurities is reduced [DEM 05].

The fundamental idea behind this component is to use a two-dimensional gas (2D-gas) of electrons circulating in a lightly doped material and resulting from the occupation of the energy levels of the potential well characteristic of a heterojunction as a conductive channel of a field transistor. Three electrodes commonly called source, gate and drain enable us to control the current and the voltage of the transistor operation. The source and drain contacts are ohmic type, and the gate contact is Schottky type. For power amplification applications, the common source arrangement is used, the source electrode being connected to ground. The gate enables us to control the current density circulating in the transistor by acting electro-statically on the electron gas; it is the low power control electrode. The drain enables us to control the operating voltage of the component; it is the high-power control electrode.

HEMT is characterized by a particular structure that consists of several layers of different materials. A silicon carbide (SiC) layer is used as a substrate on which the component is grown, a layer of nucleation that serves to achieve a tuning of the mesh between the substrate and GaN, a layer of GaN that contains the 2DEG in its upper part, AlGaN serves to create a heterojunction with the layer of GaN and other layers that provide other functions [NIF 17]. A literary study concluded that HEMT has several structures, as shown in Figure 4.1.

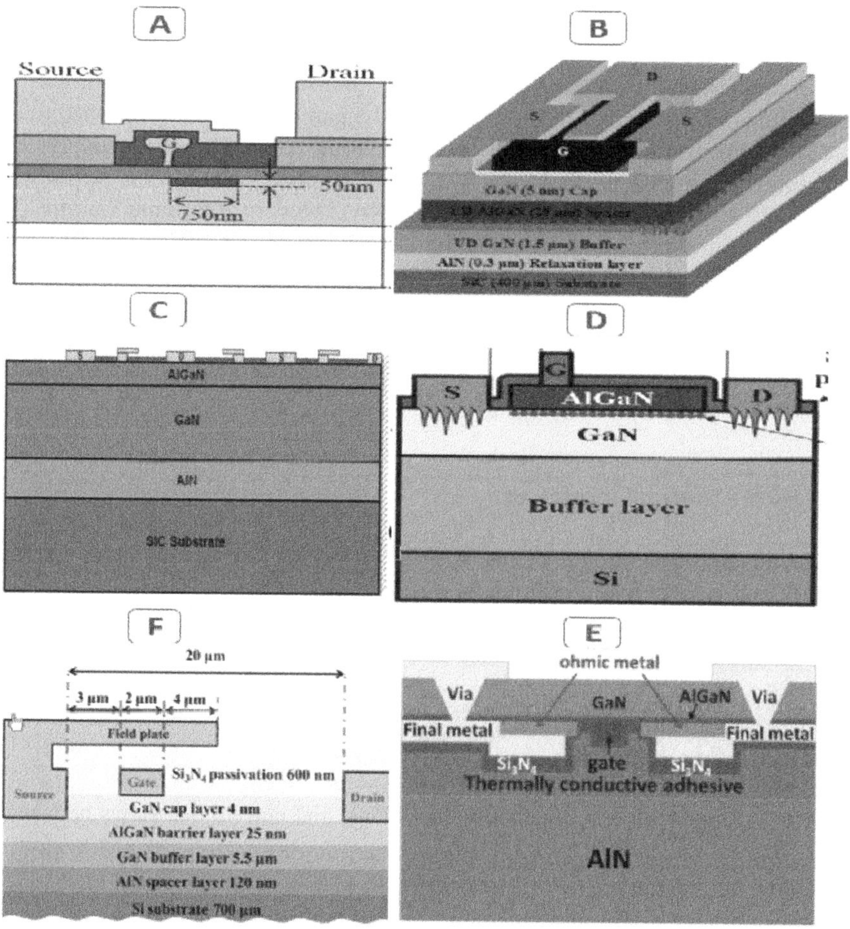

Figure 4.1. *Different HEMT structures*

It is clear that there is a difference between these structures regarding the topology and materials used; all this leads us to conclude that the high-mobility transistor is a technology that is still developing and improving in order to find the best HEMT structure to increase performance and reliability.

The HEMT has very relevant electrical and thermal characteristics. Because of these characteristics, it becomes the best candidate to be used across different systems; in airborne systems, we find these transistors at the level of its high-power amplifiers (HPA). These amplifiers are based on HEMT and can operate at very particular conditions such as high frequencies and temperatures [BAC 15]. Recently, professionals in the automotive field are beginning to use HEMTs in hybrid vehicles and motor drive circuits, following on tests and experiments that have shown the performance and advantages of these transistors over others. In the telecommunication field, HEMT is found to be particularly well positioned to meet the performance, cost structure, manufacturing capability and supply chain flexibility requirements of 4G LTE and 5G wireless base station infrastructure [IUC 19], with its low noise amplifiers (LNA). Modern electronic warfare (EW) systems require increasingly demanding characteristics. An increasing demand for power, bandwidth and efficiency in added power has led manufacturers to turn to new technologies, such as HEMT, which allows the amplifiers of these systems to ensure the operation of the bands at 6–18 GHZ [MOU 11]. This technology is not limited to the civil field, but it is also found in military applications and equipment.

4.3. HEMT thermal modeling

In this section, we are interested in the study of the thermal behavior of the HEMT; in this framework, during its operation this transistor dissipates a power at the level of the active area at the gate output; the dissipated power is a function of the voltage applied to the terminals and the current that flows in this transistor, such as:

$$P_{diss} = V_{ds}.I_{ds} \quad [4.1]$$

where V_{ds} and I_{ds} are, respectively, the voltage and current that flows between the drain and the source. To define the thermal behavior of the transistor, it is necessary to study the heat transfer within the component, and this transfer is mainly by conduction; convection and radiation are negligible since they represent less than 1.5% [AUB 04]. Based on this mode of heat exchange, the heat equation will be [ELH 15b]:

$$k.\nabla^2 T = \rho \cdot C_{P.} \frac{\partial T}{\partial t} + Q \quad [4.2]$$

where:

- Q: power dissipated in joules;
- K: thermal conductivity ($W \cdot m^{-1} \cdot K^{-1}$);
- ρ: density ($kg \cdot m^{-3}$);
- C_p: thermal mass capacity ($J \cdot kg^{-1} \cdot K^{-1}$);
- T: temperature (K).

This equation enables us to describe the thermal behavior of the transistor; for a better modeling, we will use the FEM, which will enable us to solve this problem using the Comsol Multiphysics software [COM 20].

The thermal modeling of the HEMT is developed in a 3D model using Comsol Multiphysics software, in order to observe the influence of the power dissipated at the level of the active area of the transistor during its operation, and that of the reference temperature at the substrate. The system is composed of several materials, whose properties are shown in Table 4.1 [HEM 13].

Materials	**Volume density, ρ [kg/m^{-3}]**	**Thermal conductivity KTH [W/m/K]**	**Thermal mass capacity, C_p [J/kg/K]**
Au	19,300	310	137
SiN	3,300	10	713
AlGaN	5,470	$25.\left(\frac{293}{273+T}\right)^{-1.35}$	548
GaN	6,100	$161.\left(\frac{293}{273+T}\right)^{-1.45}$	490
SiC	3,220	$416.\left(\frac{293}{273+T}\right)^{-1.5}$	690

Table 4.1. *Material properties*

Using these properties, numerical simulations of the HEMT were carried out; then the variation of the operating temperature as a function of the power dissipated P_{diss} and the reference temperature T_{ref} was studied; these results are shown in Figure 4.2. From this chart, it can be seen that the maximum temperature of the

HEMT increases following a nonlinear evolution with the increase in dissipated power and the reference temperature; it reaches critical values up to 500°C, a reference temperature of 200°C and a dissipated power density of 10 W/mm.

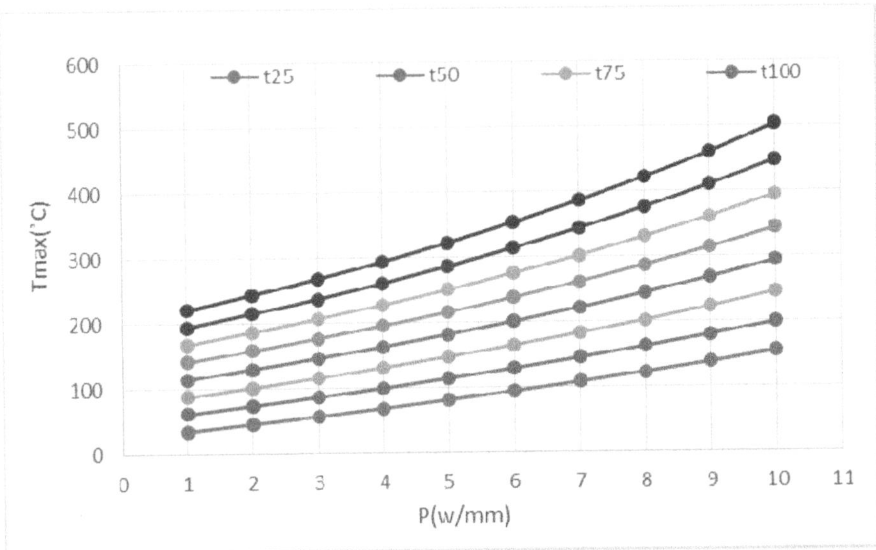

Figure 4.2. *Evolution in the operating temperature of the HEMT*

4.4. Reliability methods

The best-known definition is that by AFNOR: "Ability of a device to perform a required function under given conditions over a given period of time. The term is used as a characteristic designating a probability of success or a percentage of success". The notion of reliability is linked to several statistical and probabilistic parameters that we will deal with in the following sections.

4.4.1. *Reliability study*

The reliability analysis is done by calculating the probability of failure P_f:

$$P_f = P(G(X) \leq 0) = \int_{G(X)\leq 0} f_X(x)dx \quad [4.3]$$

where G(X) represents the performance function, which changes its value as a function of the random variable X, such that:

– G(X) > 0 defines the safety domain of the structure;

– G(X) < 0 defines the failure domain of the structure;

– G(X) = 0 defines the limit state surface.

The reliability R of the structure is calculated from expression [4.3] as follows:

$$R = 1 - P_f \tag{4.4}$$

The calculation of the probability of failure or reliability requires the calculation of the integral found in expression [4.3], but this calculation is analytically difficult and often impossible; for this reason we will need other methods to evaluate R and P_f [ELH 13].

4.4.2. *Calculating the probability of failure*

Among the methods for calculating the probabilities of failures, we have the calculation for the reliability index. FORM and SORM are two very effective methods for calculating reliability. They are based on the search for the most probable failure point (MPFP), whose knowledge is the basis of the approximations of the probability of failure in these methods. SORM and FORM enable us to link the reliability index to the probability of failure.

4.4.2.1. *Reliability index*

Calculating the probability of failure from a reliability index is a method for estimating the probability of failure from the calculation of a parameter that is called the reliability index β, for which there are several proposed reliability indices; however, the best known are Cornell and Hasofer–Lind reliability indexes.

Cornell reliability index β_c: it is obtained from the mean μ_G and the standard deviation σ_G, by the following equation:

$$\beta_C = \frac{\mu_G}{\sigma_G} \tag{4.5}$$

The major drawback of this index is that it gives different values for different expressions of the limit state function.

Hasofer–Lind reliability index β_{HL}: to remove the risk of dependence of the reliability index on the boundary state, Hasofer and Lind suggested performing the calculation of β in the space of reduced normal random variables, centered and statistically independent. To do this, the random vector X is transformed into the random vector U, such that:

$$U_i = T(X_i) \quad [4.6]$$

The random variables follow a reduced centered normal distribution and whatever $i \neq j$, Ui and Uj are mutually independent. This probabilistic transformation T requires knowledge of the statistical distributions for each of the random variables. The limit state function comes after the transformation:

$$H(U) = G[X(U)] \quad [4.7]$$

The probability of failure is thus equal to:

$$P_f = P(X(U) \leq 0) \quad [4.8]$$

Therefore:

$$P_f = \int_{H(u)\leq 0} \Phi_m(u)du \quad [4.9]$$

Φ_m is the density function of the reduced centered multinormal distribution. β_{HL} is defined as the Euclidean distance from the origin of the standard normal space to the boundary state surface $H(u) = 0$. u is a realization of the random vector U, that is, $u = (u_1, u_2, \ldots, u_m)^T$. $H(u)$ is a realization of the random variable $H(u)$. Therefore, we need to solve the following constrained optimization problem:

$$\begin{cases} \beta_{HL} = \min \sqrt{u^T \cdot u} \\ u \in R^m: \\ H(u) = 0 \end{cases} \quad [4.10]$$

This problem is equivalent to:

$$\begin{cases} \beta_{HL} = \min f(u) \\ u \in R^m: \\ H(u) = 0 \end{cases} \quad [4.11]$$

$$f(u) = \frac{1}{2} u^T \cdot u \quad [4.12]$$

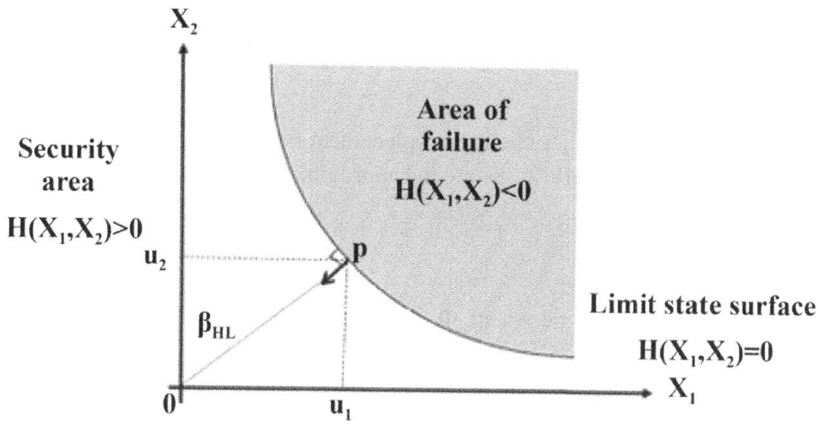

Figure 4.3. *Geometric representation of β_{HL} for a problem with two random variables*

The reliability index is therefore the minimum of the function f under the constraint $H(u) = 0$. The analytical calculation of β_{HL} is only feasible for a limited number of special cases and a numerical solution is usually required. The objective function f is a quadratic, convex form; two times continuously differentiable with respect to the variables ui. In contrast, the constraint function $H(u)$ is rarely convex. Moreover, the $H(u)$ function is sometimes complex, implicit and not continuously differentiable with respect to the variables ui.

Let u^* be the solution vector of the problem [2.4], and P^* is the point on the limit state surface such that $OP^* = u^* \cdot P^*$ is called the design point or most likely failure point (see Figure 4.4). Furthermore:

$$u^* = -\beta_{HL}\alpha \qquad [4.13]$$

α is the normal vector to the surface $H(u) = 0$ at the point u^* or its normalized gradient. The objective is then to relate β_{HL} to the probability of failure of the structure. Several methods exist and provide a more or less fine value of P_f [ELH 13].

4.4.2.2. *Approximation methods*

The most experienced approximation methods are the FORM/SORM, which are very efficient for reliability calculation. These methods are based on the search for the design point, also called the most probable failure point (MPFP), whose knowledge is the basis of the approximations for the probability of failure in the framework of these methods.

FORM: its principle is based on the replacement of the limit state surface by the hyperplane tangent to the MPFP, where the approximation of P_f becomes:

$$P_f = \Phi(-\beta) \quad [4.14]$$

where β is the reliability index and Φ is the centered reduced normal distribution function.

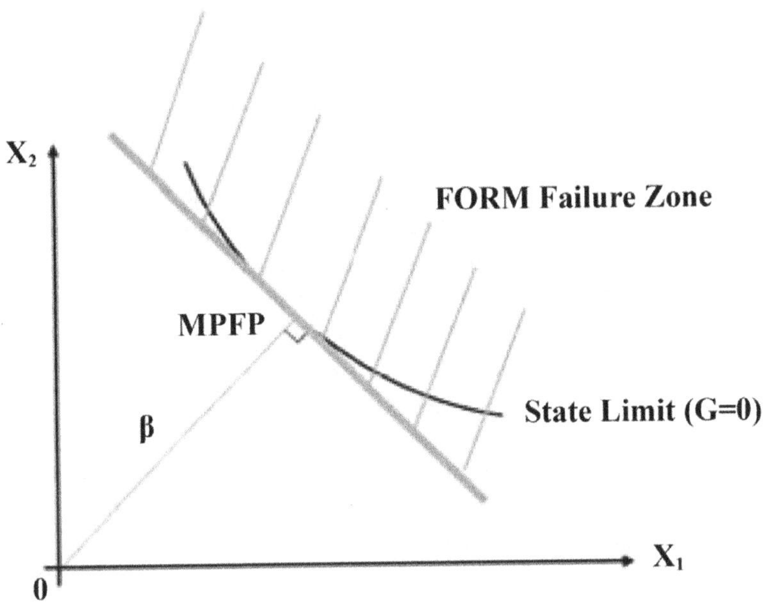

Figure 4.4. *Principle of the FORM*

SORM: If the limit state function is highly nonlinear, the approximation of the limit state surface by a hyperplane can lead to an erroneous probability of failure. It consists of replacing the limit state surface with a quadratic surface tangent and having the same curvatures of the real surface at the MPFP.

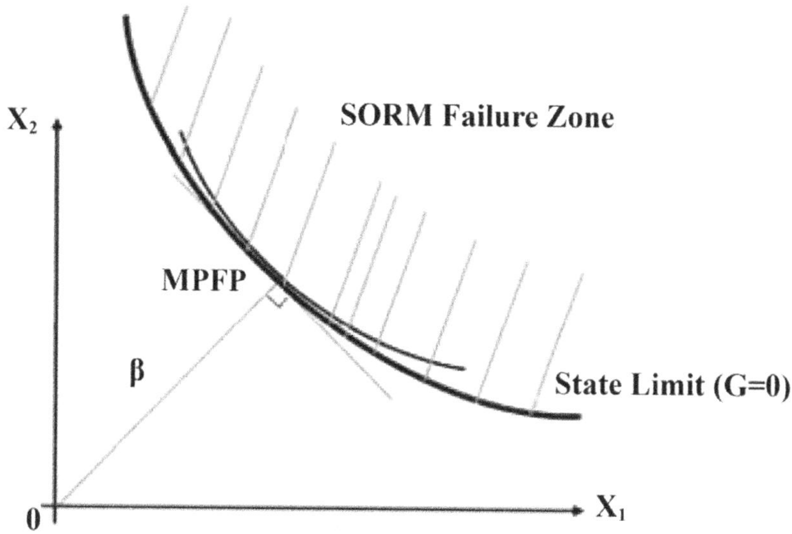

Figure 4.5. *Principle of the SORM*

SORM gives a better estimate of the probability of failure. Hence the approximation P_f has as expression:

$$P_f = \Phi(-\beta)\left(\prod_{i=1}^{n-1} \frac{1}{\sqrt{1+\beta k_i}}\right) \quad [4.15]$$

where β is the reliability index, and k_i are the principal curvatures of the G function at MPFP.

4.5. Thermo-reliability coupling

On the one hand, the thermal model developed ensures the link between the input data and the output variables; on the other hand, the reliability model takes care of the reliability calculation, due to a thermo-reliability coupling between the two software Comsol Multiphysics and Matlab. This coupling will enable us to calculate the thermal reliability of the HEMT, as per the flowchart shown in Figure 4.6.

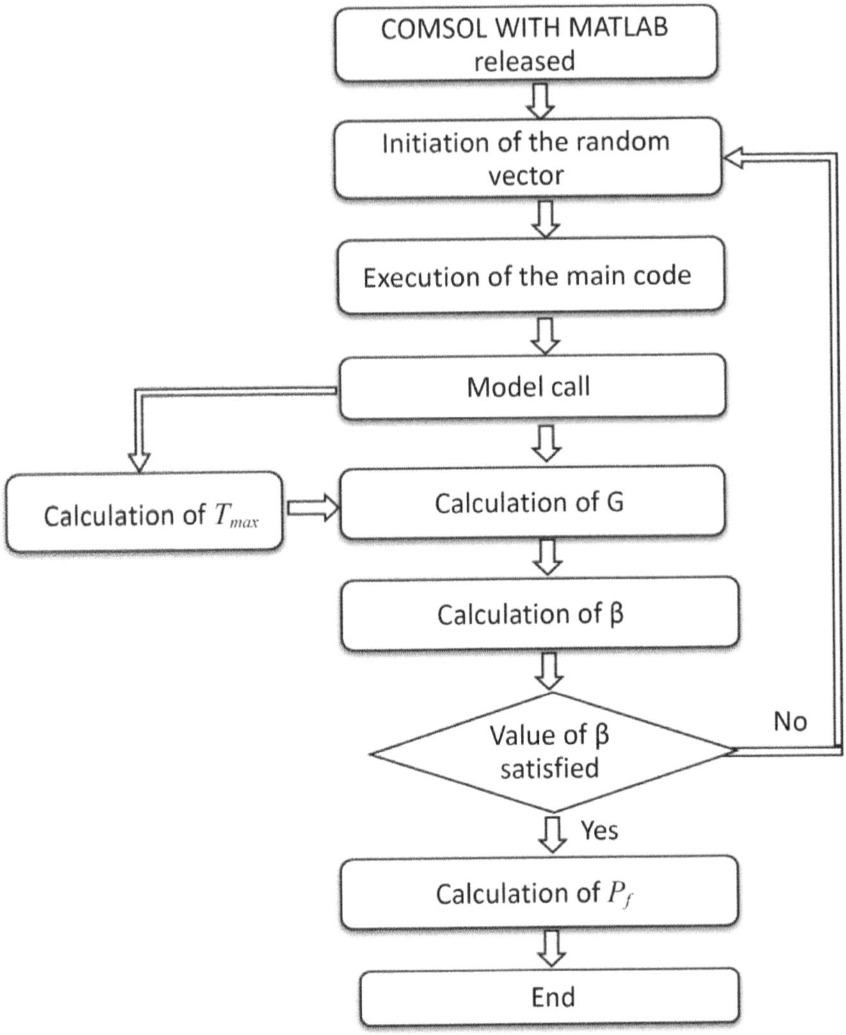

Figure 4.6. *Flowchart for thermo-fiabilistic coupling*

4.6. Calculating HEMT reliability

Thermal modeling has identified the parameters that have a great impact on the operating temperature, such as the power dissipation P_{diss} and the reference temperature T_{ref}, for which we consider that these two variables follow a normal law. If T_{max} is the maximum temperature that the system must not exceed, and T is the maximum temperature calculated by the finite element model, the performance function is written as:

$$G = T_{max} - T \qquad [4.16]$$

Using thermo-reliability coupling, the reliability index β, failure probability and reliability R of the transistor were calculated by both the SORM and FORM approximation methods. The results are given in Table 4.2.

Parameters	FORM	SORM
μ_{Pdiss} (w/mm)	5.5	5.5
σ_{Pdiss} (w/mm)	0.55	0.55
μ_{Tref} (°C)	112.5	112.5
σ_{Tref} (°C)	11.25	11.25
β	4.51	4.53
P_f	3.09 e-6	2.918 e-6
R	9.9997	9.9997
Temps (s)	2,970	3,073

Table 4.2. *Results of the reliability analysis of the HEMT*

4.7. Conclusion

In this chapter, we used the SORM and FORM to estimate the reliability index and the probability of failure for the HEMT, using the coupling between the thermal model, in finite elements, developed using the software Comsol Multiphysics, which enabled us to observe the influence of P_{diss} and T_{ref} on the operating temperature and the reliability model coded under Matlab, which ensures the probabilistic calculation of the reliability. The results obtained by the two methods are generally satisfactory as well as the calculation time. In the future, we intend to propose efficient methods based on metamodels for the calculation of reliability.

4.8. References

[ABO 11a] ABO AL-KHEER A., EID M., AOUES Y. et al., "Theoretical analysis of the spatial variability in tillage forces for fatigue analysis of tillage machines", *Journal of Terramechanics*, vol. 48, no. 4, pp. 285–295, 2011.

[ABO 11b] ABO AL-KHEER A., KHARMANDA M.G., EL HAMI A. et al., "Estimating the variability of tillage forces on a chisel plough shank by modeling the variability of tillage system parameters", *Computers and Electronics in Agriculture*, vol. 78, no. 1, pp. 61–70, 2011.

[ABO 11c] ABO AL-KHEER A., EL-HAMI A., KHARMANDA M.G. et al., "Reliability-based design for soil tillage machines", *Journal of Terramechanics*, vol. 48, no. 1, pp. 57–64, 2011.

[AMA 19] AMAR A., RADI B., EL HAMI A., "La modélisation thermique de transistor à haute puissance de type HEMT", *Incertitudes et fiabilité des systèmes multiphysiques*, vol. 3, pp. 1–7, 2019.

[AUB 04] AUBRY R., Etude des aspects électrothermiques de la filière HEMT AlGaN/GaN pour application de puissance hyperfréquence, Université des Sciences et Technologiques de Lille, 2004.

[BAC 15] BACZKOWSKI L., JACQUET J. C., JARDEL O. et al., "Temperature measurements in RF operating conditions of AlGaN/GaN HEMTs using IR microscopy and Raman spectroscopy", *European Microwave Week*, vol. 5, pp. 152–155, 2015.

[BEN 13] BEN ABDESSALEM A., PAGNACCO E., EL-HAMI A., "Increasing the stability of T-shape tube hydroforming process under stochastic framework", *International Journal of Advanced Manufacturing Technology*, vol. 69, nos. 5–8, pp. 1343–1357, 2013.

[BEN 15] BEN ABDESSALEM A., EL HAMI A., "A probabilistic approach for optimising hydroformed structures using local surrogate models to control failures", *International Journal of Mechanical Sciences*, vol. 96–97, pp. 143–162, 2015.

[BOU 96] BOUYEKHF R., EL MOUDNI A., EL HAMI A. et al., "Reduced order modelling of two-time-scale discrete non-linear systems", *Journal of the Franklin Institute*, vol. 333, no. 4, pp. 499–512, 1996.

[BOU 01] BOUYEKHF R., EL HAMI A., EL MOUDNI A., "Optimal control of a particular class of singularly perturbed nonlinear discrete-time systems", *IEEE Transactions on Automatic Control*, vol. 46, no. 7, pp. 1097–1101, 2001.

[CHE 08] CHEROUAT A., RADI B., EL HAMI A., "The frictional contact of the composite fabric's shaping", *Acta Mechanica*, vol. 199, nos. 1–4, pp. 29–41, 2008.

[COM 20] COMSOL, COMSOL multiphysics modeling software®, available at: https://www.comsol.fr/, 2020.

[DAM 17] DAMMAK K., EL HAMI A., KOUBAA S. et al., "Reliability based design optimization of coupled acoustic-structure system using generalized polynomial chaos", *International Journal of Mechanical Sciences*, vol. 134, pp. 75–84, 2017.

[DAM 19] DAMMAK K., KOUBAA S., EL HAMI A. et al., "Numerical modelling of vibro-acoustic problem in presence of uncertainty: application to a vehicle cabin", *Applied Acoustics*, vol. 144, pp. 113–123, 2019.

[DEM 05] DE MEYER S., BERGEAULT E., DE JAEGER R. J. C. et al., Etude d'une nouvelle filière de composants HEMTs sur technologie nitrure de gallium. Conception d'une architecture flip-chip d'amplificateur distribué de puissance à très large bande, University of Limoges, 2005.

[ELH 93] EL HAMI A., LALLEMENT G., MINOTTI P. et al., "Methods that combine finite group theory with component mode synthesis in the analysis of repetitive structures", *Computers and Structures*, vol. 48, no. 6, pp. 975–982, 1993.

[ELH 96] EL HAMI A., RADI B., "Some decomposition methods in the analysis of repetitive structures", *Computers and Structures*, vol. 58, no. 5, pp. 973–980, 1996.

[ELH 09] EL HAMI A., RADI B., CHEROUAT A., "Treatment of the composite fabric's shaping using a Lagrangian formulation", *Mathematical and Computer Modelling*, vol. 49, nos. 7–8, pp. 1337–1349, 2009.

[ELH 11] EL HAMI A., RADI B., "Comparison study of different reliability-based design optimization approaches", *Advanced Materials Research*, vol. 274, pp. 113–121, 2011.

[ELH 13] EL HAMI A., RADI B., *Incertitudes, optimisation et fiabilité des structures*, Lavoisier, Paris, 2013.

[ELH 15a] EL HAMI A., POUGNET P., *Embedded Mechatronic Systems 1; Analysis of Failures, Predictive Reliability*, ISTE Press, London, and Elsevier, Oxford, 2015.

[ELH 15b] EL HAMI A., POUGNET P., *Embedded Mechatronic Systems 2: Analysis of Failures, Modeling, Simulation and Optimization*, ISTE Press, London, and Elsevier, Oxford, 2015.

[GUE 15a] GUERINE A., EL HAMI A., WALHA L. et al., "Approach for the dynamic analysis of one stage gear system with uncertain parameters", *Mechanism and Machine Theory*, vol. 92, pp. 113–126, 2015.

[GUE 15b] GUERINE A., EL HAMI A., FAKHFAKH T. et al., "A polynomial chaos method to the analysis of the dynamic behavior of spur gear system", *Structural Engineering and Mechanics*, vol. 53, no. 4, pp. 819–831, 2015.

[GUE 16a] GUERINE A., EL HAMI A., WALHA L. et al., "A polynomial chaos method for the analysis of the dynamic behavior of uncertain gear friction system", *European Journal of Mechanics, A/Solids*, vol. 59, pp. 76–84, 2016.

[GUE 16b] GUERINE A., HAMI A.E., WALHA L. et al., "Dynamic response of a Spur gear system with uncertain friction coefficient", *Advances in Engineering Software*, vol. 120, pp. 45–54, 2016.

[GUE 17] GUERINE A., EL HAMI A., WALHA L. et al., "Dynamic response of wind turbine gear system with uncertain-but-bounded parameters using interval analysis method", *Renewable Energy*, vol. 113, pp. 679–687, 2017.

[HEM 13] HEMTS G., MARCON D., MENEGHESSO G. et al., *Reliability Analysis of Permanent Degradations*, vol. 60, no. 10, pp. 3132–3141, 2013.

[HUA 16] HUANG C., RADI B., EL HAMI A., "Uncertainty analysis of deep drawing using surrogate model based probabilistic method", *Manufacturing Technology*, vol. 86, nos. 9–12, pp. 3229–3240, 2016.

[IUC 19] IUCOLANO F., BOLES T., GaN-on-Si HEMTs for wireless base stations, *Materials Science in Semiconductor Processing*, vol. 98, Elsevier Ltd, Amsterdam, 2019.

[JAN 17] JANNOUN M., AOUES Y., PAGNACCO E. et al., 'Probabilistic fatigue damage estimation of embedded electronic solder joints under random vibration", *Microelectronics Reliability*, vol. 78, pp. 249–257, 2017.

[KHA 14] KHARMANDA G., IBRAHIM M.-H., ABO AL-KHEER A. et al., "Reliability-based design optimization of shank chisel plough using optimum safety factor strategy", *Computers and Electronics in Agriculture*, vol. 109, pp. 162–171, 2014.

[MAB 17] MABROUK I.B., EL HAMI A., WALHA L. et al., "Dynamic vibrations in wind energy systems: application to vertical axis wind turbine", *Mechanical Systems and Signal Processing*, vol. 85, pp. 396–414, 2017.

[MAK 16] MAKHLOUFI A., AOUES Y., EL HAMI A., "Reliability based design optimization of wire bonding in power microelectronic devices", *Microsystem Technologies*, vol. 22, no. 12, pp. 2737–2748, 2016.

[MOR 02] MORO T., EL HAMI A., EL MOUDNI A., "Reliability analysis of a mechanical contact between deformable solids", *Probabilistic Engineering Mechanics*, vol. 17, no. 3, pp. 227–232, 2002.

[MOU 11] MOUGINOT G., Potentialités des transistors HEMTs AlGaN-GaN pour l'amplification large bande de fréquence ; effets limitatifs et modélisation bande de fréquence ; effets limitatifs et modélisation, University of Limoges, 2011.

[NIF 17] NIFA I., LEROUX C., TORRES A. et al., "Characterization of 2DEG in AlGaN/GaN heterostructure by Hall effect", *Microelectronic Engineering*, vol. 178, pp. 128–131, 2017.

[RAD 07] RADI B., EL HAMI A., "Reliability analysis of the metal forming process", *Mathematical and Computer Modelling*, vol. 45, nos. 3–4, pp. 431–439, 2007.

[SAR 11] SARSRI D., AZRAR L., JEBBOURI A. et al., "Component mode synthesis and polynomial chaos expansions for stochastic frequency functions of large linear FE models", *Computers and Structures*, vol. 89, nos. 3–4, pp. 346–356, 2011.

5

Warranty Cost

5.1. Introduction

5.1.1. *The evolution of the warranty*

Reliability is the performance of a system in executing its functions for a given duration, under specific conditions of use, and associated with the probability of "survival". In other words, from the design of vehicles, we accept that a "small" portion of the production will not be functional or will fail. In order to protect the consumer, governments and car manufacturers very quickly proposed warranty clauses. This notion has appeared since the birth of motor vehicles in the history of humankind.

For example, in 1930, American market rules appeared with constraints on vehicle warranties [BOG 30] such as 90 days on batteries (already at that time!). It was not until 1975 that a federal law, the Magnuson & Moss Act [MAG 75], also known as the Lemon law, imposed consumer protection on manufacturers for newly purchased vehicles. This was the first time that a 1-year warranty was imposed, with the possibility of extending it to 9 years. Since the 1990s, manufacturers have understood the marketing importance of the warranty. This is why they offer specific warranties for the complete vehicle or for certain systems or components. It was during this period that China proposed a consumer protection law [CHI 93], but without any particular constraints on car manufacturers. Certain texts attempted to regulate consumer protection in 1994 (European Directive 93/13/EEC), but it was only in 1999 that the European Commission legislates [EUR 99] on certain aspects of the sale and warranties of consumer goods by imposing, for the first time,

For a color version of all the figures in this chapter, see http://www.iste.co.uk/elhami/applied3.zip.

Chapter written by David DELAUX.

a 2-year warranty. This clause includes all car sales in the European territory. More recently, on October 1, 2013, China enacted a brand-new law, the San Bao Law, imposing 3 years or 60,000 km of warranty on new vehicles. This demonstrates a legal will to control and amplify the warranty of vehicles by the state. This relatively slow amplification has been accelerated by the car manufacturers themselves in order to offer better services to their customers.

Figure 5.1 gives a summary of the warranty conditions and their evolution.

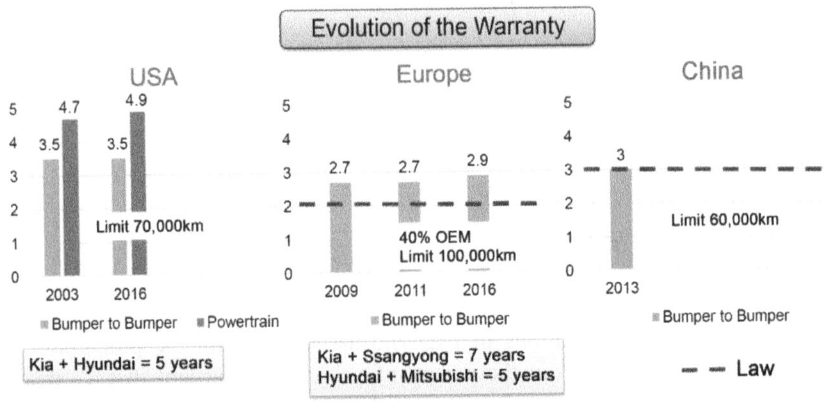

Figure 5.1. *Evolution of the automobile warranty across the world*

Figure 5.1 shows that car manufacturers go above and beyond the legal requirements for warranty.

All car manufacturers worldwide and their suppliers therefore have financial data that is more or less accessible. Often this data is considered "highly sensitive", with access limited to top management (Finance Department, Quality Department). For the United States, part of the warranty cost information is publicly accessible.

5.1.2. *The warranty cost*

In 2001, the flagship of the American oil industry, ENRON, went bankrupt within hours. The visibility of transactions and risk provisions was not communicated, which triggered a panic on Wall Street. This is why, in 2002, the federal Financial Accounting Standards Board (FASB) imposed the obligation to disclose, at least once a year, these financial results, including costs and provisions for warranty risks. Therefore, for almost 20 years, a very large amount of financial data has been made available.

Figure 5.2 shows a compilation of U.S. manufacturers, warranty costs from 2002 to 2015.

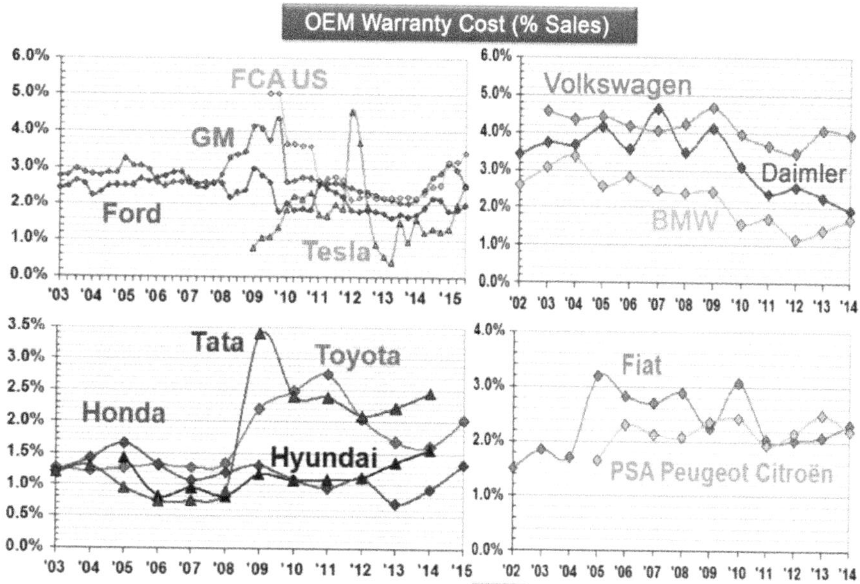

Figure 5.2. *Evolution of warranty costs in the United States*

From the data set in Figure 5.2 (more than 150 points), it is possible to statistically model the variable "warranty cost", the relative cost as compared to turnover, by considering it as a random variable of a stochastic process.

We have the shape of this law in Figure 5.3 representing a log-normal law whose probability density function is in the form of:

$$f(x;\mu,\sigma) = \frac{1}{x\sigma\sqrt{2\pi}} e^{\left(-\frac{(\ln(x)-\mu)^2}{\sigma^2}\right)} \qquad [5.1]$$

The average relative warranty cost for all U.S. automakers is 2.15% of sales. This cost can be considered the warranty "cushion", a random "normal" cost over the production year.

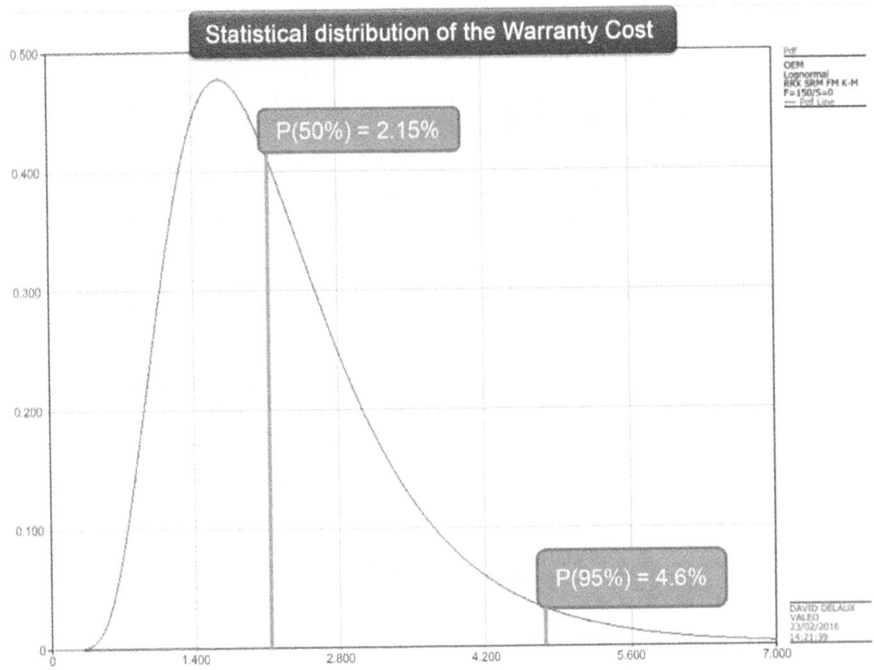

Figure 5.3. *Statistical distribution of the warranty cost (% sales)*

If we take the average price of a vehicle worldwide to be 20,000€, associated with a world production of 80 million cars, the average annual global warranty cost would be 34.3 billion euros. According to a German study [BEA 07], the annual cost of warranty would be more pessimistic in the order of 40–44 billion euros. It is also necessary to take into account the warranty "crises", which could represent a cost of around 4.6% (or more) of the turnover.

This introduction demonstrates that there is a wealth of information on warranty costs for the automotive world. Since warranty costs can reflect the quantity of defective products, it would be tempting to find a mathematical model to explore this wealth of information, by interpreting the reliability performance of the system. However, it should be pointed out that the set of defects constituting the warranty cost is not only due to a design problem. The study [WU 11] shows that more than 10% of warranty claims are explained by misused system or various human factors. The entropy of warranty return information is very important (i.e. lack of rigor in dealer and garage networks).

Nevertheless, this entropy can be minimized by considering the warranty cost as a whole, and as a large amount of data. In other words, we can consider the warranty cost as an indicator of the system reliability performance by a holistic approach.[1]

5.2. Warranty and reliability

5.2.1. *Qualitative analysis*

A qualitative analysis of warranty problems in the automotive industry, which can lead to "fleet recalls", allows us to understand the root causes of the defect and the countermeasures to be applied in order to stop the financial hemorrhage.

According to the Japanese Ministry of Transport, the study of automobile "fleet recalls" from 2006 to 2010 [JAP 07] states that 54% of the claims are due to design reliability defects with the following breakdown:

– 47% related to insufficient test criteria;

– 5% related to product development;

– 2% related to lack of comprehensive vehicle testing.

Such studies highlight that a qualitative analysis is able to extract points directly linked to the reliability engineering of the mechatronic system. We can extract three axes of improvement:

– the importance of implementing reliability engineering throughout the product lifecycle to achieve long-term goals and reduce warranty costs;

– an in-depth analysis of the failure mode that occurred during the warranty period will allow the same defect to be related to the different reliability engineering phases. This will accelerate the resolution of the warranty problem;

– an analysis of the set of parts and data collected in the field (publicly accessible) will give a perspective toward the recalibration of reliability actions for the next development programs. This will contribute to either an increase in system reliability or "just-in-time" design.

The more innovative and complex the mechatronic system, the more important the monitoring and analysis of the warranty information should be.

1 Holistic: In epistemology or in the human sciences, relating to the doctrine that brings the knowledge of the particular, of the individual to that of the whole, of the whole in which it fits (Larousse).

5.2.2. *Quantitative analysis*

Warranty returns, when they exist, provide a great deal of information to extract laws from, such as:

– law of driving (number of km driven per year);

– storage law (number of months before sale of the vehicle);

– law of failure (per month of commissioning or per kilometer driven).

It is therefore possible to apply mathematical models such as a failure rate (also called hazard function) in order to assess the life of the system in terms of its real-world usage. By exploiting the data of warranty returns, it is possible to have access to the duration of use (in months, in hours, etc.) or to the mileage associated with the time of use between its sale and the observation of the failure. These elements are used to calculate the operational reliability of the system. Several analytical models can be used, but one of the more famous is the Weibull distribution.

The probability density function takes the following form:

$$f(T) = \frac{\beta}{\eta}\left(\frac{T-\gamma}{\eta}\right)^{\beta-1} e^{-\left(\frac{T-\gamma}{\eta}\right)^{\beta}} \quad [5.2]$$

The Weibull distribution is very versatile, which gives this model great flexibility in its handling and application. This distribution is most often used in the form of the reliability function, denoted as R(t).

$$R(t) = e^{-\left(\frac{T-\gamma}{\eta}\right)^{\beta}} \quad [5.3]$$

Figure 5.4 shows a reliability function of a mechatronic system.

In fact, it is not surprising that this statistical model is still widely used in the automotive world, certainly due to the fact that Professor Waloddi Weibull worked for a key player in the industry, SAAB, a major car manufacturer. He was not working on developing a theoretical mathematical model but rather trying to solve concrete physics problems, such as strength of material dilemmas and fatigue of materials [LAN 08]. This demonstrates the true nature of reliability engineering, which is a multidisciplinary science in the service of industry.

In the qualitative analysis of warranty returns data, there are two types of failures:

– “minor” defects which have a low impact on the driver and are often related to driving comfort (i.e. a defect in the ceiling light, a window regulator button, etc.);

– “major” defects limiting the driving of the vehicle (i.e. noise, engine in degraded mode, etc.).

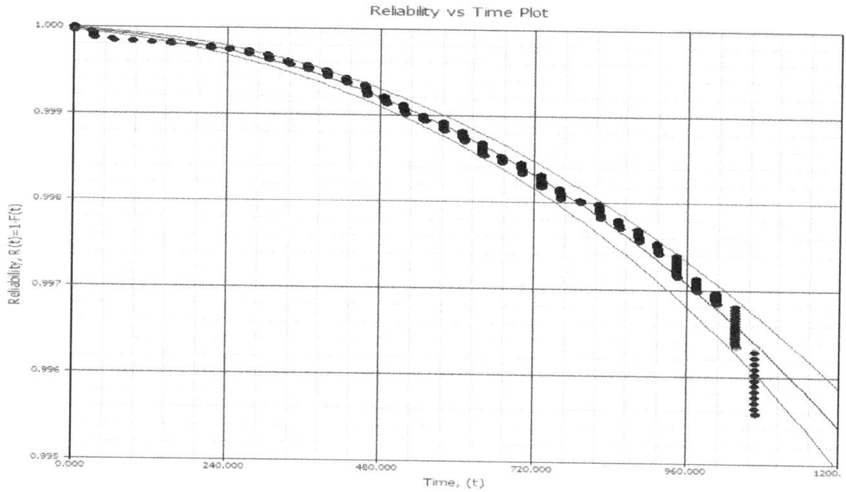

Figure 5.4. *Reliability function of a mechatronic system*

It becomes difficult, even impossible, to follow up on the reliability of elements responsible for causing “minor” defects. However, even in the case of “major” defects, it is necessary for the car manufacturer to control the flow of defective parts throughout the network (dealer, garage, dealers, etc.).

The reliability performance can nevertheless be estimated from the warranty data due to statistical models.

5.3. Reliability estimation models

5.3.1. *Parametric, non-parametric and other models*

Several studies propose the use of Poisson distributions [KAL 91], which estimate warranty costs as a function of time. Moskowitz and Chun [MOS 94] suggest using a variant of the Poisson model to predict the quantity of warranty returns by taking into account the dimensions of the warranty (year and km) by fitting the parameter λ with the functions of time and mileage. In the same vein, Hu and Lawless [HU 96] advocate a technique for modeling warranty costs by truncating the data and assuming that warranty returns follow a Poisson process. Oh and Bai [OH 01] present a method arguing for the use of a parametric model based on product observations that exceed the warranty period.

Another model proposes to use the cumulative time and quantity of warranty return per month in service, a model developed by Wasserman [WAS 92] and Robinson-McDonald [ROB 91]. Their approaches are based on a non-parametric model to statistically establish "$R(t)$-claims" per thousand cumulative vehicles per month in service.

Wasserman [WAS 92] also developed a linear and dynamic predictive model of the $R(t)$ function using model vehicle data from different production years. Using a log-log paper, Robinson and McDonald [ROB 91] suggest plotting the points of $R(t)$ and finding the best line by linear regression. Singpurwalla and Wilson [SIN 93] developed a variant of the failure model for failures in warranty returns as a function of time and mileage. They derive the two major failure distributions and present a method for predicting $R(t)$ using an asymmetric log-log analytical analysis.

5.3.2. *Mixed models*

The mechatronic systems of the vehicle interact with each other and influence each other inexorably. This leads to complexity for analysis by mixed models. Figure 5.5 illustrates this singularity.

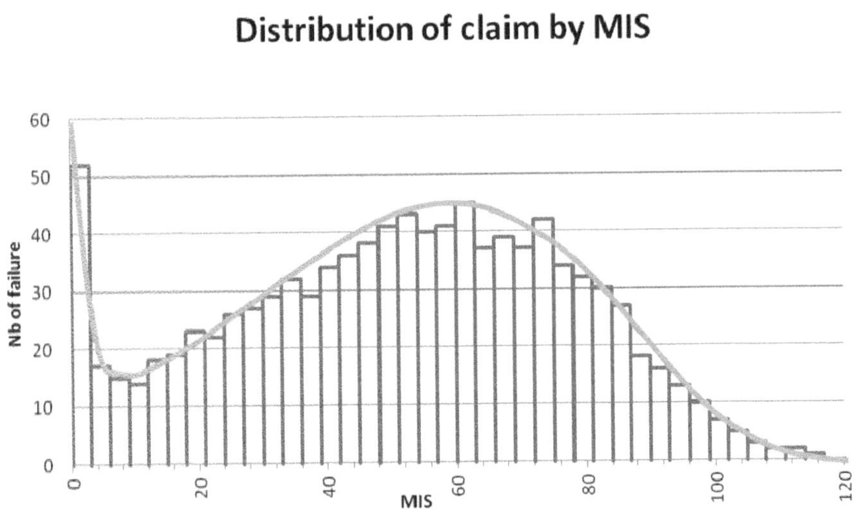

Figure 5.5. *Distribution of mixed reliability*

It is complicated to separate the failure rates, as shown in Figure 5.5, because in fact this curve is the result of the sum of all the system failure rates observed in warranty returns. This can be illustrated in Figure 5.6.

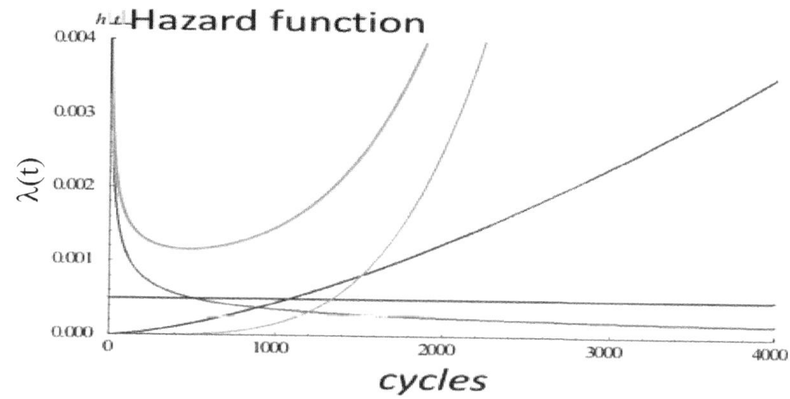

Figure 5.6. *Multiple failure rates*

It would be possible to decompose the warranty return data by failure mode to discover the contributors of "early", "random" or "wear-and-tear" failures (the three parts of the bathtub curve in Figure 5.9).

These different failure modes are illustrated in Figure 5.7.

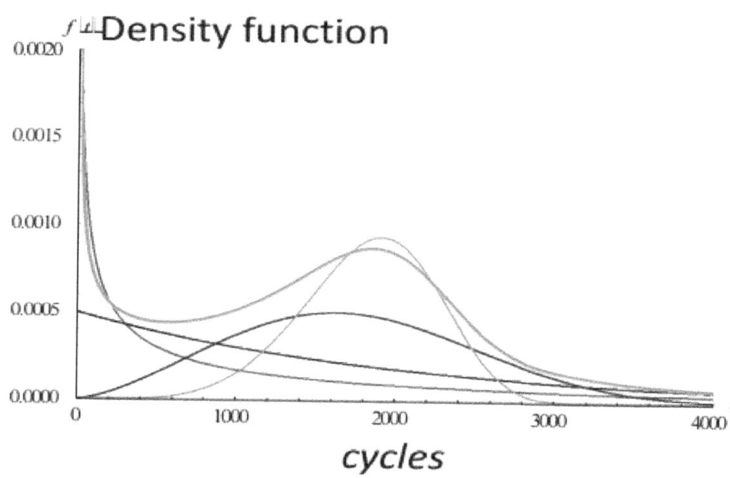

Figure 5.7. *Different failure modes*

Chung et al. [CHU 86] and Meeker [MEE 87] are interested in a mixed Weibull model, equivalent to:

$$F(t) = \begin{cases} p\left(1 - e^{-(\alpha t)^{\beta}}\right), 0 \leq t < \infty \\ 1 \qquad\qquad\qquad\quad , \quad t = \infty \end{cases} \quad [5.4]$$

This model can be applied to mechatronic systems whose failures are "electronic" (components, PCBA, etc.). The authors highlight the need for a mixed model in order to provide intuitive arguments regarding the system's lifetime properties.

This approach is also supported by regular observation of warranty on the ground, in terms of two competing system failure modes. Lifetime estimation can follow the model demonstrated by Mori et al. [MOR 91]. Their approach implies that multiple failure modes occur more often than a multitude of "early" failure rates. Majeska [MAJ 03] proposes a combination of uniform and Weibull distributions capable of distinguishing two failure modes:

$$F(t) = \begin{cases} p\theta & , \quad t = 0 \\ p\theta + p(1-\theta)\frac{t}{t^*} + (1-p)\left(1 - e^{-\alpha t^{\beta}}\right) & , 0 < t \leq t^* \\ p + (1-p)\left(1 - e^{-\alpha t^{\beta}}\right) & , \quad t > t^* \end{cases} \quad [5.5]$$

5.3.3. *Advantages and disadvantages*

The different statistical models (non-exhaustive list) elaborate probabilistic simulations of lifetimes from warranty feedback data.

Depending on the complexity of the model, we can appreciate:

– pre-delivery defects;

– multiple failure modes;

– various design improvements.

Such analyses require a significant amount of data (often estimated at more than 1,000 parts) to have a stable statistical picture of the reliability (or unreliability in the case of warranty returns).

Nevertheless, the reliability expert exploiting the operational data (warranty) must have a thorough knowledge of the mechatronic system in order to identify all the parameters of the statistical model being considered. The use of sophisticated software tools and the mastery of mathematical models are also essential.

In addition to the above constraints, we must add two important shortcomings.

Firstly, it is necessary to have a large number of defective parts collected in warranty to have a statistical model. Secondly, it is difficult or impossible to have a projection for the reliability of the system at 15 years (or 300,000 km) with a statistical model using information in a relatively short warranty period (often from 1–3 years).

Based on the principles of parametric, non-parametric and mixed models, we propose below a new mathematical model based on warranty return costs.

5.4. New models for estimating reliability from warranty costs

There are attempts to estimate reliability on the basis of financial data [BAR 14], but it can only be applied to a purely electronic system and under certain conditions. As for the model proposed by Rahman [RAH 06], it is mainly interested not in reliability but rather in the problem of extending the warranty.

5.4.1. *Assumptions*

Two strong assumptions form the basis of this new approach as follows:

The warranty cost $\omega_{\%}$ related to the turnover is *homogeneous with a failure rate.*

We propose this formulation for the relative warranty cost:

$$\omega_{\%} = \frac{n.C_{\omega}}{N.C_S} \qquad [5.6]$$

where:

– C_{ω}: cost of the warranty invoice (in euros);

– C_S: cost of selling a new product (in euros);

– n: the number of defective parts within the warranty period;

– N: the number of parts produced during the warranty period.

Note that the cost of the invoice must be weighted by the cost of labor when changing the part, as well as a financial "penalty" factor that may be imposed by the car manufacturer on its direct supplier.

In other words, the factor C_ω takes the following analytical form:

$$C_\omega = C_s + (C_l.\alpha) \qquad [5.7]$$

where:

C_l: labor cost for part change (in euros);

α: coefficient depending on the warranty contract.

According to a study of a German car manufacturer's database (+50,000 warranty data), the following rule is proposed:

– warranty contract less than 3 years: $\alpha = 1.1$ (default);

– warranty contract between 3 and 5 years: $\alpha = 1.3$;

– warranty contract longer than 5 years: $\alpha = 1.4$.

The analysis of the database shows that the average invoice in Europe is 263€. It is interesting to note that the cost of the invoice is between two and six times the cost of the new part.

n and *N* are inaccessible unknowns, but assuming we know the cost parameters (C_l, C_s, C_w), the relative cost of the warranty $\omega_{\%}$ is homogeneous with the failure rate.

The used market *price* (more precisely the discount of a new vehicle) is representative of *unreliability*.

The quotation of a used vehicle can be the image of the confidence in the reliability of the vehicle. Indeed, we can imagine that we want to sell our vehicle quickly if we have encountered breakages or if we consider that we have reached the limit of its lifespan.

This vision is very simplistic, considering the great complexity of the second-hand market that depends on many marketing parameters (the make and model of the vehicle, the geographical location, etc.). This hypothesis does not work either for top-of-the-range models (such as Aston Martin, Ferrari or Maserati) or collector models (such as the 1960 Jaguar E-Type) since the price of the second-hand vehicle depends more on the image of the vehicle than on its performance over time.

Despite the above considerations, it is possible to apply this assumption for the vast majority of cases affected by the warranty cost.

Therefore, after the purchase of a new vehicle, the price decreases dramatically in the first 3 years (with an average loss of about 30–50% of the new price), which may represent the first phase of the bathtub curve (Figure 5.9): the "early failures". From 3 years onwards and for a certain period of time, the used car price decreases monotonically, which could represent the "random failure" phase of the bathtub curve. Finally, after a significant number of years of use, the price drops significantly, which is considered as the beginning of the "wear-and-tear" phase of the bathtub curve.

Figure 5.8 illustrates the case of a Honda Civic in Spain (Torrevieja) [VEH 17].

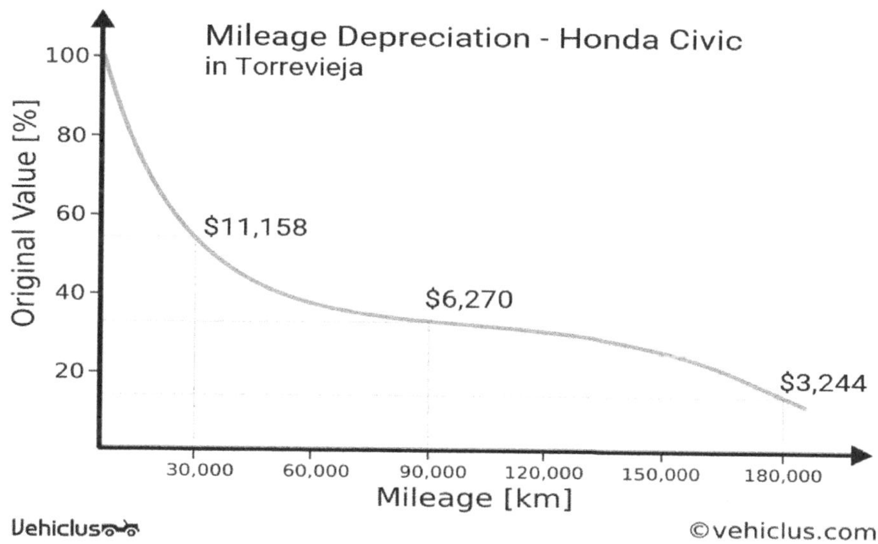

Figure 5.8. *Depreciation of the price of a Honda Civic in Spain*

5.4.2. Definition of the transition between "random" and "wear-and-tear" phases

The evolution of the failure rate of a mechatronic system follows the main phases of the bathtub curve, recalled here in Figure 5.9.

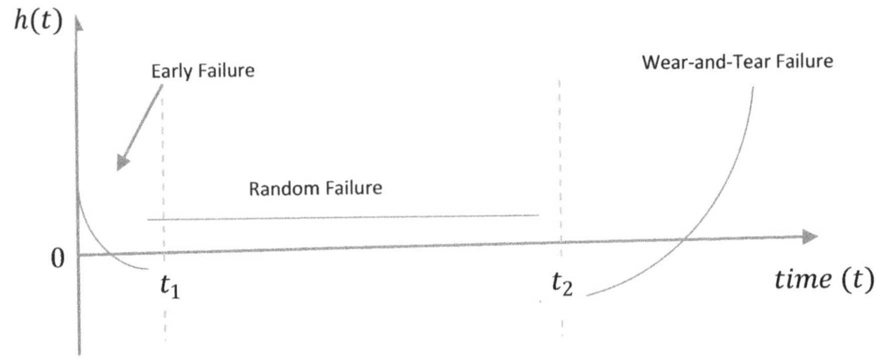

Figure 5.9. *Bathtub curve*

The delimitation between "random" and "wear-and-tear" failures is provided by the time t_2. The whole question is to approach this time in a realistic way. In order to do this, we use hypothesis 2, explained above, by applying it to two dimensions of the warranty: temporal (years) and mileage (kilometers).

a) Kilometric sizing (km): several studies of the used car market allow us to locate the mileage horizon to be around 160,000 km.

We have selected three studies to illustrate this pivot point:

– Longhow Lam's study on 2000 Renault Clio in the Netherlands (see Figure 5.10) [LON 19];

– Mats Olsen and M. S. Hernandez's study of 300 vehicle models from Norway, Spain and Sweden (see Figure 5.11) [OLS 16];

– Reddit's study of 335 models of Toyota Prius and Honda Fit in the United States (see Figure 5.12) [RED 18].

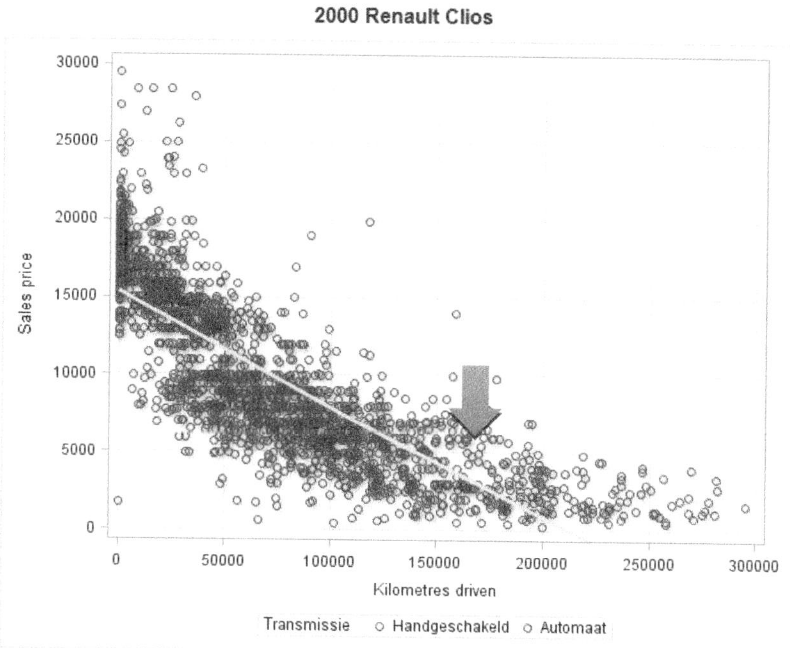

- Pivot point at approximately 160,000 km

Figure 5.10. *Renault Clio depreciation (Lam study)*

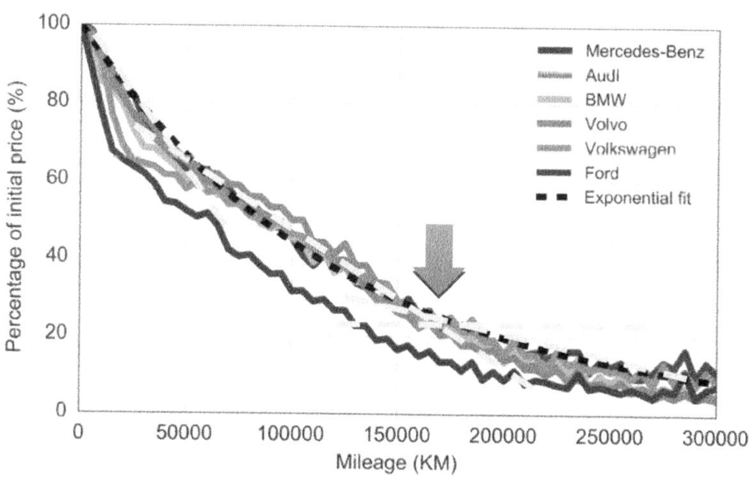

- Pivot point at approximately 160,000 km

Figure 5.11. *Depreciation of six automotive brands (Olsen study)*

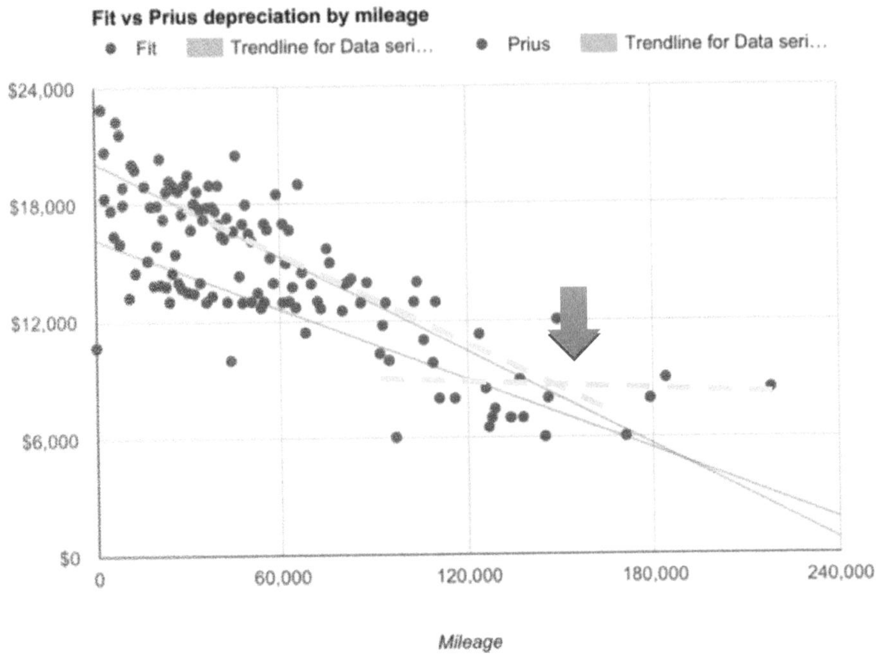

- Pivot point at approximately 160,000 km

Figure 5.12. *Toyota versus Honda depreciation in the United States (Reddit study)*

b) Temporal sizing (years).

Three studies are presented below to illustrate this point:

– Vehiclus Study of the Audi Q7 model in China (Shanghai) between 2002 and 2017 (see Figure 5.13) [VEH 19];

– Mats Olsen, M. S. Hernandez (see Figure 5.14) [OLS 16];

– Free By 50, a study of the American market (see Figure 5.14) [FRE 13].

Figure 5.13 shows a singular point around 7 years.

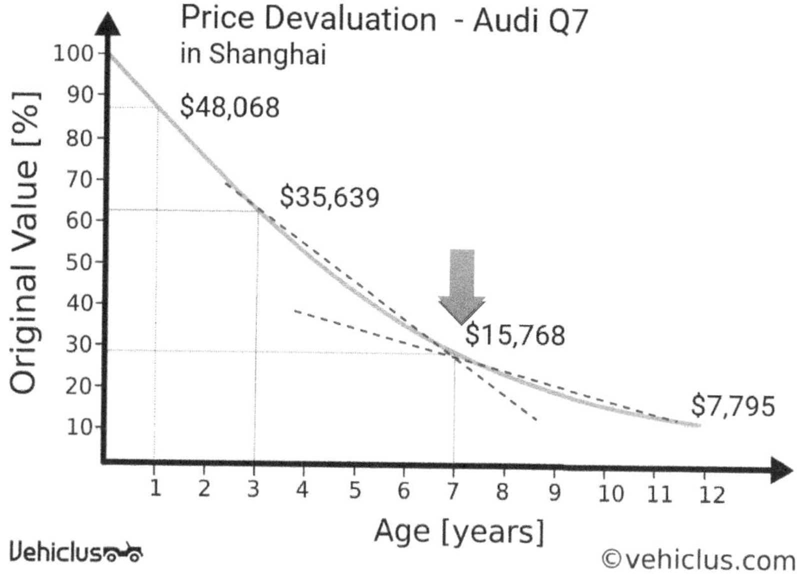

Figure 5.13. *Depreciation Audi Q7 – Shanghai (China – Vehiclus study)*

Figure 5.14 appears to show a pivot point around 7–8 years.

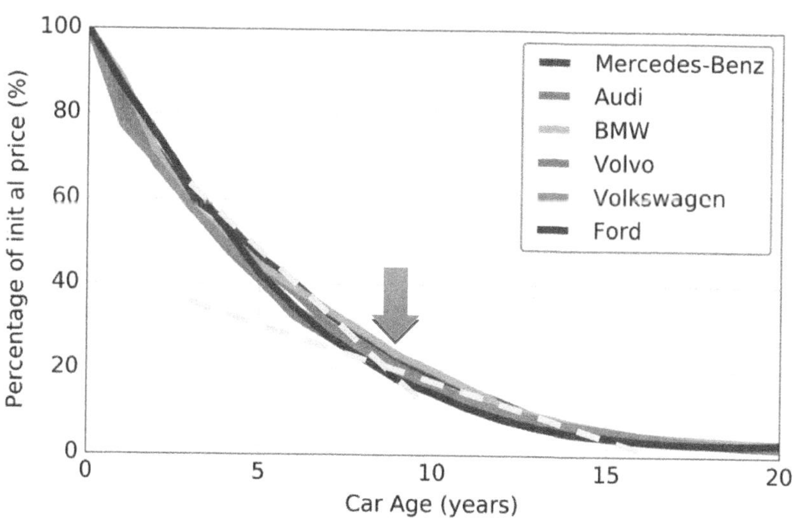

Figure 5.14. *Depreciation of European vehicles (Olsen study)*

Figure 5.15 would indicate a change of around 8 years.

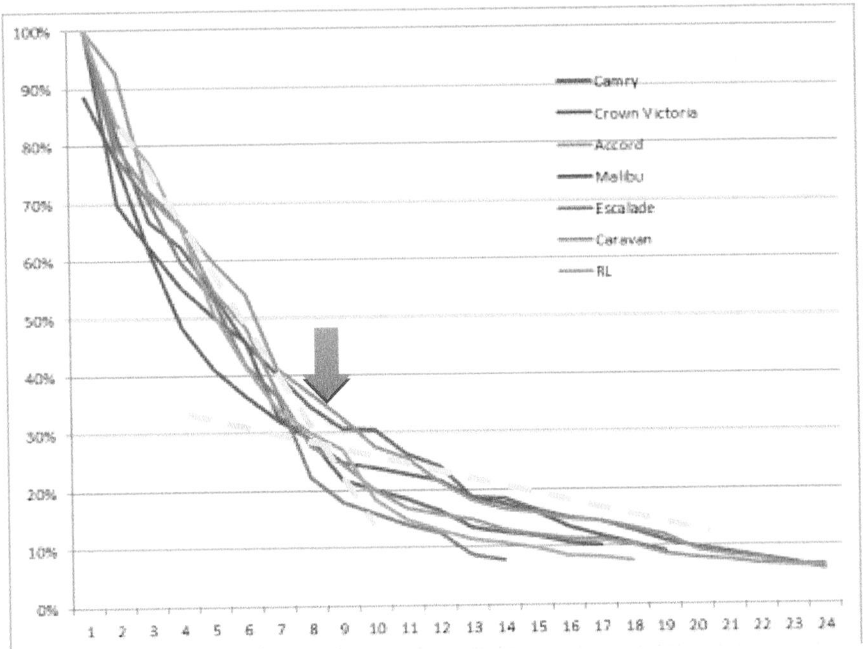

Figure 5.15. *Depreciation of American vehicles (FreeBy50 study)*

In 2018, an interview was conducted with car manufacturers to assess the French, German and Japanese market. For obvious reasons of confidentiality, the names will not be mentioned.

In summary, this study is condensed in Table 5.1:

Market	Considering the start of the "wear-and-tear" phase on the vehicle
French	Between 5 and 7 years
German	After 8 years
Japanese	Around 10 years

Table 5.1. *Beginning of wear-and-tear phase according to market*

In summary, it seems coherent to converge toward an inflection point of 8 years.

The 8 years or 160,000 km horizon is also consistent with a generic driving law [NHT 06] estimated at 20,000 km/year.

5.4.3. *New operational reliability model for the "random" phase*

The "random" or "constant failure rate" phase is defined for the range $t < t_2$; $t_2 \in$ [3 years; 8 years].

The failure model in this phase follows a Poisson process [YAN 07], which takes the following form:

$$R(t) = e^{(-\lambda . t)} \qquad [5.8]$$

Let us transpose the parameter $\lambda(t)$ as a function of the relative warranty cost $\omega_{\%}$:

$$\lambda(t) = \omega_{\%} . \delta \qquad [5.9]$$

where the warranty cost parameter δ is:

$$\delta = \frac{C_S}{C_S + (C_l . \alpha)} \qquad [5.10]$$

By injecting [5.10] and [5.9] into [5.8], we obtain the failure probability $F(t)$:

$$F(t) = 1 - R(t) = 1 - e^{(-\lambda t)} = 1 - \exp\left\{-\omega_{\%} * \left(\frac{\overline{C_S}}{\overline{C_S} + (\overline{C_l} \times \alpha)}\right)\right\}(t) \qquad [5.11]$$

5.4.4. *New operational reliability model for the "wear-and-tear" phase*

The "wear-and-tear" phase, or increasing failure rate, is therefore defined over the range $t > t_2$; $t_2 = 8$ years.

The failure model in this phase follows a power law process that can take the form of a two-parameter Weibull distribution [YAN 07] of the type:

$$R(t) = e\left[-\left(\frac{t}{\eta}\right)^{\beta}\right] \qquad [5.12]$$

where β: shape parameter must be greater than 1. Conservatively, it is possible to set $\beta = 2$ (log-normal law) as a safe value.

The question now is to define the scaling parameter η.

Placing ourselves at $t = t_2$, just at the junction of the two phenomena, "random" and "wear-and-tear", we have the following equality:

$$R(t_2) = e^{(-\lambda . t_2)} = e\left[-\left(\frac{t_2}{\eta}\right)^{\beta}\right] \qquad [5.13]$$

Deriving η by:

$$\eta = \frac{t_2}{[\lambda . t_2]^{\frac{1}{\beta}}} \qquad [5.14]$$

Integrating [5.14] into [5.13], we obtain the reliability model for the "wear-and-tear" phase:

$$\mathrm{R(t)} = \exp\left\{-\frac{\mathrm{t}}{\mathrm{t_2}} . (\omega_{\%} . \mathrm{t_2}) . \left(\frac{\overline{\mathrm{C_S}}}{\overline{\mathrm{C_S}} + (\overline{\mathrm{C_l}} . \alpha)}\right)^{\frac{1}{\beta}}\right\}^{\beta} \qquad [5.15]$$

where:

– $\overline{\mathrm{C_S}}$: estimate for the price of the new part;

– $\overline{\mathrm{C_l}}$: estimate for the labor cost to replace the defective part.

5.5. Applied automotive case studies

The assumptions are as follows:

Variable	Description	Value
t_2	Time horizon of transition from "random" to "wear-and-tear" failure mode	8 years
$\omega_{\%}$	Relative warranty cost as a percentage of sales	0.50%
$\overline{C_s}$	Estimated price of a new mechatronic system	€30.00
$\overline{C_l}$	Estimated cost of a repair (labor part)	€120.00
α	Penalty coefficient according to the warranty conditions 3 years or 100,000 km	1.1
β	Weibull shape parameter (safe value)	2

Table 5.2. *Automobile application assumptions for warranty cost*

1) "Random" phase: the estimate for the probability of failure in the "random" phase between 3 years (end of warranty) and 8 years (beginning of the "wear-and-tear" phase) is defined by Table 5.1.

The numerical application of equation [5.11] with the assumptions outlined in Table 5.2 yields the reliability projection $R(t)$:

$$R(t < 8\,years) = 99.907\% \qquad [5.16]$$

In the so-called "random" part of the system's life, we more often express the probability of failure expressed in ppm (parts per million). In other words, equation [5.16] is expressed in $F(t)$ as:

$$F(t < 8\,years) = 1 - R(t) = 925\text{ ppm}$$

One property of the failure rate in this phase is that it is constant. It is therefore possible to say that during the period of 3–8 years of production, it is likely to have 925 ppm defective parts.

If the production mass is 100,000 parts per year, or 500,000 parts between 3 and 8 years of production, we will therefore have 463 cumulative defective systems.

Note that the parameter of the relative warranty cost $\omega_{\%}$ has an important and proportional weight in the result. For example, if we have four times more warranty costs ($\omega_{\%} = 2\%$), we will have four times more defective parts (1,848 failures).

2) "Wear-and-tear" phase: the reliability estimate in the "wear-and-tear" phase is set at the end-of-life design goal of 15 years.

Numerical application of equation [5.15] alongside the assumptions outlined in Table 5.2 yields the reliability projection $R(t)$:

$$R(t = 15\text{ years}) \sim 72\%$$

If we had four times more warranty costs ($\omega_{\%} = 2\%$), the reliability performance drops to 52%.

Note that often the reliability target of car manufacturers at 15 years is around 95% (for a non-safety system). In this case, we must have a relatively "slow" wear-and-tear mechanism driven by a β that is close to 1.5.

5.6. Conclusion

In the automotive world, as in many other industries, the warranty is a fundamental building block of operational reliability. Not only because it is a crucial financial commitment that can bring a company to ruin, but also because it is a wealth of information that can be used as an indicator of reliability in the real world.

Numerous statistical models were developed very early on in the history of the automobile industry, becoming more and more sophisticated over time. However, the models presented in this chapter provide a simple and original way with which to explore the richness of the sometimes-public information of financial risks through the warranty cost to sales ratio. This chapter suggests safe zones by which to conservatively assess operational reliability performance in two significant periods of the life of a mechatronic system: the "random phase" (between 3 years and 8 years) and the "wear-and-tear phase" (beyond 8 years). With a Bayesian approach, we can use the knowledge of the reliability assumed by the warranty cost to project up to 15 years of life.

The models will be all the more efficient if we integrate knowledge of the field (mission profile, rolling law, etc.), the failure mechanisms in warranty (refinement of the shape parameter β) and the exploration of a large number of data (part cost, repair bill, potential financial penalty, etc.).

These models can be criticized in terms of the strong assumptions made. However, Isaac Newton's postulate, stated more than 330 years ago, still holds true and is applicable here: "In experimental philosophy, propositions gathered from phenomena by induction should be considered either exactly or very nearly true notwithstanding any contrary hypotheses, until yet other phenomena make such propositions either more exact or liable to exceptions".

5.7. References

[ABR 64] ABRAMOWITZ M., STEGUN I.A., *Handbook of Mathematical Functions, Applied Mathematics Series*, vol. 55, National Bureau of Standards, Washington, 1964 (reprinted 1968 by Dover Publications, New York).

[ADV 16] ADVANCED INDUSTRIES, "Automotive revolution – perspective towards 2030", McKinsey & Company, 2016.

[AFN 50] AFNOR, Démonstration de la tenue aux environnements – Conception et réalisation des essais en environnement, NFX 50-144-1 to 6, 2016.

[AIT 57] AITCHISON J., BROWN J.A.C., *The Lognormal Distribution*, Cambridge University Press, Cambridge, 1957.

[AME 16] AMERICA DEPARTMENT OF DEFENSE, Highly accelerated life testing and highly accelerated stress screening methodology, 6-F80B601, 2016.

[ASH 97] ASHLEY S., "Getting a hold on mechatronics", *Mechanical Engineering-CIME*, vol. 119, pp. 5, 1997.

[AST 09] ASTE, Guide d'application de la démarche de personnalisation en environnement mécanique, PRO NORMDEF 0101 Edition 01, Ministère de la Défense, available at: http://www.aste.asso.fr/fr/pag-488138-Guide-climatique-et-mecanique.html, 2009.

[AUT 13b] AUTO FOREVER, Production mondiale automobile depuis 1900 (automobile production since 1900), available at: http://www.auto-forever.com/production-automobile-depuis-1900/, 2013.

[BAH 15] BAHRET A., "The origins, philosophy, methods and practices of HALT", *ASTR Conference*, Cambridge, 9–11 September 2015.

[BAR 14] BARTHOLDT M., "Determination of target reliability levels based on value to the customer and warranty budgets", *Probabilistic Safety Assessment and Management PSAM 12*, Honolulu, Hawaii, 2014.

[BEA 07] BEARINGPOINT, Global automotive warranty survey report, Report, 2007.

[BER 03] BERTRAM T., BEKES F., GREUL R., et al., "Modelling and simulation for mechatronic design in automotive systems", *Control Engineering Practice*, vol. 11, pp. 179–190, 2003.

[BOG 30] BOGART G.G., FINK E.E., "Business practice regarding warranties in the sale of goods", *Illinois Law Review*, vol. 25, pp. 400–417, 1930.

[BOI 00] BOITEUX D., "Essais d'endurance de systèmes embarqués", *PSA*, 2000.

[BON 77] BONNET D., LALANNE C., "Choix des essais - Analyse de l'environnement mécanique vibratoire réel en vue de l'élaboration de spécifications d'essais", *Journées ASTE*, vol. 5, 1977.

[BON 15] BONATO M., DELAUX D., "Synthesis and validation of accelerated vibration durability tests", *RAMS*, pp. 1–6, 2015.

[BOR 02] BORNER M., STRAKY H., WEISPFENNING T., et al., "Model based fault detection of vehicle suspension and hydraulic brake systems", *Mechatronics*, vol. 12, no. 8, pp. 999–1010, 2002.

[BRA 16] BRASSEUL J., *Petite Histoire des faits économiques – De nos origines à nos jours*, Armand Colin, 2016.

[BRO 01] BROUSSELY M., HERREYRE S., BIENSAN P., et al., "Aging mechanism in Li ion cells and calendar life predictions", *Journal of Power Sources*, vols. 97–98, pp. 13–21, 2001.

[BYU 17] BYUN J.E., NOH H.M., SONG J., *Reliability Growth Analysis of k-out-of-N Systems using Matrix-based System Reliability Method*, Elsevier, 2017.

[CAM 08] CAMANHO P.P., DAVILA C.G., PINHO S.T., et al., *Mechanical Response of Composites*, Springer, 2008.

[CEE 09] CEEES, Reliability for a mature product from the beginning of useful life, no. 9, 2009.

[CHU 86] CHUNG M.K., WU S.H., HERRIN G.D., "The use of a mixed Weibull model in occupational injury analysis", *J. Occupational Accidents*, vol. 7, pp. 239–250, 1986.

[COM 94] COMERFORD R., "Mecha...what?", *IEEE Spectrum*, vol. 31, no. 8, pp. 46–49, 1994.

[DEL 06] DELAUX D., "Reliability validation of engine cooling modules with a tailoring tests of vibration, thermal shock and pressure pulsation", *Revue Essai & Simulation*, 2006.

[DEL 17] DELAUX D., EL HAMI A., GRZESKOWIAK H., *Reliability of High Power Mechatronic Systems 2*, ISTE Press, London and Elsevier, Oxford, 2017.

[DEM 04] DEMMOU H., KHALFAOUI S., GUILHEM E., et al., "Critical scenarios derivation methodology for mechatronic systems", *Reliability Engineering and System Safety*, vol. 84, pp. 33–44, 2004.

[DES 96] DESJARDIN L., "A day in the life of mechatronic engineers 10 years from now", *SAE International Congress and Exposition*, no. SAE96C038, Detroit/Michigan, 1996.

[DEU 18] DEULOFEU PIQUET J., DEULOFEU BATLLORI R., *Pascal Le fondateur de la théorie des probabilités, Génies des mathématiques*, RBA Barcelona, 2018.

[DEV 20] DEVALAN, P., "Introduction à la mécatronique", *Techniques de l'ingénieur*, available at : https://www.techniques-ingenieur.fr/base-documentaire/mecanique-th7/mecatronique-42509210/introduction-a-la-mecatronique-bm8000/, 2020.

[DEW 86] DE WINNE J., "Equivalence of fatigue damage caused by vibrations", *IES Proc.*, pp. 227–234, 1986.

[DIM 99] DIMARIA D.J., STATHIS J.H., "Non-Arrhenius temperature dependence of reliability in ultrathin silicon dioxide films", *Applied Physics Letters*, vol. 74, no. 12, pp.1752–1754, 1999.

[DJE 17] DJERRADA S., "Quel est le prix moyen d'une réparation ?", *Autos Infos* [Online], available at: https://www.auto-infos.fr/article/reparation-quel-est-le-prix-moyen-d-une-prestation.146484, 2017.

[DOD 93] DODGE Y., *Statistique – Dictionnaire Encyclopédique*, Dunod, Paris, 1993.

[ERR 03] ERROL M., "The fog of the war: Eleven lessons from the life of Robert S. McNamara", American documentary, 2003.

[GAU 96] GAUDEL M.-C., MARRE B., SCHLIENGER F., et al., *Précis de génie logiciel*, Masson, Paris, 1996.

[GIB 30] GIBRAT R., "Une loi des répartitions économiques : l'effet proportionnel", *Bull. Stat. Gen. Fr.*, vol. 19, pp. 469–513, 1930.

[GRI 01] GRIMHEDEN, M., HANSON, M., "What is mechatronics? Proposing a didactical approach to mechatronics", *1st Baltic Sea Workshop on Education in Mechatronics*, Kiel, Germany, 2001.

[GRI 11] GRIEU M., Etude de la fatigue des joints brasés de composants éléctroniques soumis à des sollicitations thermomécaniques, vibratoires et combinees, thesis, 2011.

[HU 96] HU X.J., LAWLESS J.F., "Estimations of rate and mean functions from truncated recurrent event data", *J Am Stat Assoc*, vol. 91, no. 433, pp. 300–310, 1996.

[IEE 16] IEEE, *Recommended Practice on Software Reliability*, 2016.

[KAL 91] KALBFLEISCH J.D., LAWLESS J.F., ROBINSON J.A., "Methods for the analysis and prediction of warranty claims", *Technometrics* vol. 33, pp. 273–285, 1991.

[KEC 03] KECECIOGLY D., SUN F.-B., Environmental Stress Screening: Its Quantification, Optimization and Management, DESTech, 2003.

[KHA 14] KHANNA K., SINGH G., ARNEJA G.P., et al., "Use of mechatronics in today's industry: a review", *Conference: National Conference on Advances in manufacturing Systems-CAMS-2014*, September 5-6 2014.

[KIM 06] KIM Y.B., NOGUCHI H., AMAGAI M., "Vibration fatigue reliability of BGA-IC package Pb-free solder and Pb Sn Solder", *Microelectronics Reliability*, vol. 46, pp. 459–466, 2006.

[KOM 17] KÖMMLING A., JAUNICH M., POURMAND P., et al., "Influence of ageing on sealability of elastomeric O-rings", *Macromolecular Symposia*, vol. 373, 2017.

[LAL 84] LALANNE C., "Maximax response and fatigue damage spectra", *The Journal of the Environmental Sciences, Part I*, vol. XXVII, no. 4, *Part II*, no. 5, pp. 40–44, 1984.

[LAL 14a] LALANNE C., *Mechanical Vibration and Shock, Volume 2: Mechanical Shock*, ISTE Ltd, London and John Wiley & Sons, New York, 2014.

[LAL 14b] LALANNE C., *Mechanical Vibration and Shock Analysis Series, Volume 5: Specification Development*, ISTE Ltd, London and John Wiley & Sons, New York, 2014.

[LAM 19] LAM L., Car depreciation Renault Clio, available at: https://longhowlam.wordpress.com/2015/04/17/car-depreciation-and-regression-splines-3/, 2019.

[LAN 08] LANNOY A., *Maîtrise des risques et sûreté de fonctionnement : Repères historiques et méthodologiques*, Tec & Doc, Paris, 2008.

[MAJ 03] MAJESKE K.D., "A mixture model for automobile warranty data", *Reliability Engineering and System Safety*, vol. 81, pp. 71–77, 2003.

[MAT 17] MATHOWORLD, Exponential integral, available at: http://mathworld.wolfram.com/ExponentialIntegral.html, 2017.

[MEE 87] MEEKER W.Q., "Limited failure population life tests: application to integrated circuit reliability", *Technometrics*, vol. 29, no. 1, pp. 51–65, 1987.

[MIL 93] MILLBANK, J., "Mecha-what!", *Mechatronics Forum Newsletter*, vol. 6, 1993.

[MOI 12] MOIR I., SEABRIDGE A., *Design and Development of Aircraft Systems*, 2nd edition, John Wiley and Sons, 2012.

[MOL 04] MOLLA J., JACOBSA J., KUSTERSB R., et al., "Defect detection oriented lifecycle modeling in complex product development", *Information and Software Technology*, vol. 46, pp. 665–675, 2004.

[MON 98] MONCELET G., Application des RdP à l'évaluation de la sûreté de fonctionnement des systèmes mécatroniques du monde automobile, PhD Thesis, LAAS, Toulouse, 1998.

[MOR 91] MORI S., ARAI N., KANEKO Y., et al., "Polyoxide thinning limitation and superior ONO interpoly dielectric of nonvolatile memory device", *IEEE Trans Electron Dev*, vol. 38, pp. 270–276, 1991.

[MOR 17] MORI Y., *Mechanical Vibrations: Applications to Equipment*, ISTE Ltd, London and John Wiley & Sons, New York, 2017.

[MOS 94] MOSKOWITZ H., CHUN Y.H., "A Poisson regression model for two-attribute warranty policies", *Naval Res Logistics*, vol. 41, pp. 355–376, 1994.

[MYR 05] MYRNA O., "Samuel Alderson, 90: inventor of dummies used to test car safety", *Los Angeles Times*, 17 February 2005.

[NEW 17] NEWCOMER J.T., BIERBAUM R.L., FRAMEWORK A., *Taxonomy for the Design and Analysis of Margins*, IEEE, 2017.

[NHT 06] NHTSA, Vehicle survivability and travel mileage schedules, Technical Report, DOT HS 809-952, 2006.

[NIE 10] NIEDERMEYER E., What's wrong with this picture: total recall edition, Recall on the rise, available at: http://www.thetruth aboutcars.com/2010/07/whats-wrong-with-this-picture-total-recall-edition/, 2010.

[OCO 12] O'CONNOR P.D.T., KLEYNER A., *Practical Reliability Engineering*, 5th edition, John Wiley and Sons, 2012.

[OH 01] OH Y.S., BAI D.S., "Field data analysis with additional after warranty failure data", *Reliab Engng Syst Safety*, vol. 72, no. 1, pp. 1–8, 2001.

[OKA 00] OKAMOTO H., *Phase Diagrams for Binary Alloys*, ASM International – The Materials Information Society, 2000.

[OLS 16] OLSEN M., HERNANDEZ M.S., Price your car with data, available at: http://bytes.schibsted.com/price-car-data/, 2016.

[ORA 00] ORAEE H., "Quantative approach to estimate the life expectancy of motor insulation systems", *IEEE Transactions on Dielectrics and Electrical Insulation*, vol. 7, no. 6, pp. 790–796, 2000.

[PEA 16] PEARSON P., CALLEN R., BLANQUART J.P., et al., *Reliability Prediction Data Sources and Methodologies for Space Applications, Guidelines for Reliability Prediction for Space Systems*, ADS-RPDSM-TN-005, 2016.

[PIE 01] PIERRAT L., "Modèle d'Arrhenius : mieux estimer le facteur d'accélération thermique", *Journal Essai Industriel*, pp. 28–30, 2001.

[PIE 13a] PIERRAT L., DELAUX D., "Analytical improvement of the stress-strength method by considering a realistic strength distribution", *Applied Reliability Symposium – Berlin*, 2013.

[PIE 13b] PIERRAT L., FEIDT M., "Introduction d'une variation aléatoire de température dans le modèle de dégradation thermodynamique d'Arrhenius", *Société Française de Thermique, congress*, 2013.

[PLA 04] PLANTE J., Alternative test methods for electronic parts, NASA Electronic Parts and Packaging (NEPP) Program, 2 Dynamic Range Corporation, 2004.

[PLO 96] PLÖGERT K., "The tailoring process in the German v-model", *Journal of Systems Architecture*, vol. 42, pp. 601–609, 1996.

[PRA 65] PRABHU N.U., *Stochastic Processes*, Macmillan, London, 1965.

[PRA 05] PRABHAKAR MURTHY D.N., BLISCHKE W.R., *Warranty Management and Product Manufacture*, Springer, 2005.

[RAH 06] RAHMAN A., CHATTOPADHAY G., "Review of long-term warranty policies", *Asia Pacific Journal of Operational Research*, 2006. DOI: 10.1142/S021759590600108X.

[RIE 93] RIEUNEAU F., "Sûreté de fonctionnement en phase de developpement des systèmes embarqués automobiles", *Integrated Logistics & Concurrent Engineering*, Montpellier, 1993.

[RIV 11] RIVET R., "Accelerated stress testing HALT & HASS", *Conference N.I.A.R.*, 2011.

[ROB 91] ROBINSON J.A., MCDONALD G.C., "Issues related to field reliability and warranty data", *Data Quality Control: Theory and Pramatics*, Marcel Dekker, New York, 1991.

[RZE 03] RZEVSKI, G., "On conceptual design of intelligent mechatronic systems", *Mechatronics*, vol. 13, no. 10, pp. 1029–1044, 2003.

[SAE 12] SAE, *Handbook for Robustness Validation of Automotive Electrical/Electronic Modules*, 2012.

[SCH 04] SCHOENIG R., Définition d'une méthodologie de conception des systèmes mécatroniques sûrs de fonctionnement, PhD Thesis, Institut National Polytechnique de Lorraine, 2004.

[SEG 99] SEGAL V., NATTRASS D., RAJ K., et al., "Accelerated thermal aging of petroleum-based ferrofluids", *Journal of Magnetism and Magnetic Materials*, vol. 201, pp. 70–72, 1999.

[SIE 05] SIEMERS C., FALSETT R., SEYER R., et al., "Reliable event-triggered systems for mechatronic applications", *Journal of Systems and Software*, vol. 77, no. 1, pp. 17–26, 2005.

[SIN 93] SINGPURWALLA N.D., WILSON S., ""The warranty problem: its statistical and game theoretic aspects", *SIAM Rev*, vol. 35, pp. 17–42, 1993.

[THE 14] THEMAVISION VÉHICULES & MOBILITÉS, Pwv Autofacts 2014, available at: http://www.them avision.fr/jcms/rw_438175/infographie-le-marche-automobile-mondial-en-2020-quelle-repartition-de-la-production, 2014.

[TOL 98] TOLLENAERE, M., *Conception de produits mécaniques*, Hermes, Paris, 1998.

[TOY 04] TOYOTA, Estimate of the number of components in a vehicle, available at: http://www.toyota.co.jp/en/kids/faq/d/01/04/, 2004.

[VEH 17] VEHICLUS, Honda Civic for sale in Torrevieja, available at: http://www.vehiclus.com/Torrevieja/Honda/Civic, 2017.

[WAS 76] WASHINGTON AND LEE LAW REVIEW, The Magnuson-Moss Warranty-Federal Trade Commission Improvement Act: Protecting Consumers Through Product Warranties, vol. 33, no. 1, art. 5, available at: https://scholarlycommons.law.wlu.edu/wlulr/vol33/iss1/5/, 1976.

[WAS 92] WASSERMAN G.S., "An application of dynamic linear models for predicting warranty claims", *Comput Indus Engng*, vol. 22, no. 1, pp. 37–47, 1992.

[WEI 51] WEIBULL W., "A statistical distribution function of wide applicability", *J. Appl. Mech. Trans. ASME*, vol. 18, no. 3, pp. 293–297, 1951.

[WIK 15] WIKI COMMONS, Unidentified rural letter carrier with modified Model-T Ford, available at: https://commons.wikimedia.org/wiki/File:Unidentified_rural_letter_carrier_with_modified_Model-T_Ford.jpg and https://commons.wikimedia.org/wiki/File:Model_T_on_a_New_Road_(8113428270).jpg, 2015.

[WIK 16] WIKI COMMONS, Motor show Geneva, available at: https://commons. wikimedia.org/wiki/File:2016-03-1_Geneva_Motor_Show_0999.JPG?uselang=fr, 2016.

[WIK 17] WIKI COMMONS, Benz Patent Motorwagen 1886, available at: https://commons.wikimedia.org/wiki/File:790751_14498 29_800_ 530_benz_patent_motorwagen_1886.jpg, 2017.

[WU 11] WU S., "Warranty claim analysis considering human factors", *Reliability Engineering and System Safety*, vol. 96, pp. 131–138, 2011.

[YAN 07] YANG G., *Life Cycle Reliability Engineering*, John Wiley & Sons, 2007.

[YAS 69] YASKAWA-ELECTRIC, C., available at: https://www.yaskawa.se/en/news-events/news/article/news/mechatronics-a-yaskawa-first/?tx_news_pi1%5Bcontroller%5D=News&tx_news_pi1%5Baction%5D=detail&cHash=422f74815a9a61f674bec1549dbd0f59 [Accessed: 2019], 1969.

[ZIE 96] ZIEGLER C., Sûreté de fonctionnement d'architectures informatiques embarquées sur automobile, PhD Thesis, LAAS, Toulouse, 1996.

[ZVE 13] ZVEI ROBUSTNESS VALIDATION WORKING GROUP, *Handbook for Robustness Validation of Semiconductor Devices in Automotive Applications*, ZVEI, Zentralverband Elektrotechnik, 2013.

6

Reliability Evaluation of a Luxury Watch Product: Application of the Stress–Strength Method to a Mechanical Component

6.1. Introduction

This chapter focuses on the field of luxury goods, and more precisely on the watch industry. The requirements and expectations of customers are increasingly high with regard to the quality of products in a competitive sector. The mechanical watch is subject to different uses depending on the country or the type of wearer. In order to guarantee both the performance of mechanical watches and customer satisfaction, it is therefore necessary to evaluate the reliability of the products.

In order to ensure the functioning of the mechanical watch for several days, these watches benefit from an energy accumulation system: the watch barrel. Like any mechanical component subjected to external stresses, the watch barrel is subjected to failure modes.

In such a context, this chapter seeks to apply the stress–strength method in order to evaluate the reliability of the barrel of a mechanical watch. This evaluation is carried out starting from the quantification of customer uses and the characterization of barrel behaviors.

For a color version of all the figures in this chapter, see http://www.iste.co.uk/elhami/applied3.zip.

Chapter written by Matthieu SALLIN and Anthony PONCET.

6.2. Presentation of the watch and its case study

6.2.1. *The mechanical watch*

The mechanical movement watch, or mechanical watch, is a type of watch commonly composed of the movement and the casing within a case. The winding to give the watch its operating energy can be done manually or automatically through the movements of the wearer.

A standard mechanical watch has about 130 components assembled in three sub-assemblies: the power source, the regulating parts and the display. The number of components is obviously much higher for watches with complications (chronograph, date, moon phases, etc.). The energy supplied by the spring is transmitted by the gear train. The escapement transmits the impulses while the balance wheel divides the time. The winding, manual or automatic, allows for the tightening of the spring.

Figure 6.1. *Diagram of a mechanical watch*

NOTES ON FIGURE 6.1.–

1) Barrel

2) Cog

3) Escapement

4) Oscillating balance

5) Winding stem/rotor

6) Analog display

6.2.2. *Case study of the barrel spring*

Our study will focus on the study of a mechanical watch with manual winding, and more precisely, on the driving force of the watch: the barrel. This consists of a gear wheel and a cylindrical case closed by a cover. The barrel turns freely on a shaft and contains the mainspring, which is attached to the drum by its outer coil and to the shaft by its inner coil.

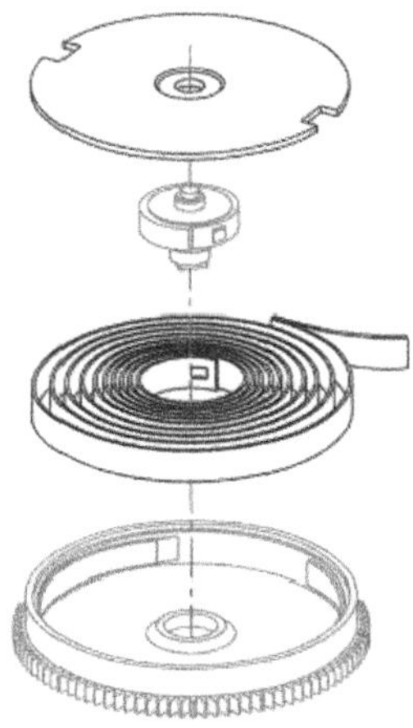

Figure 6.2. *Diagram of a barrel spring*

We shall study the behavior and lifespan of the barrel spring.

6.2.3. *Identification of failure modes and damaging factors*

The barrel is the driving force of the mechanical watch and is a component that deserves great attention. A preliminary analysis allows us to identify the probable failures of the barrel system as well as the factors damaging the system.

System failure mode	Material failure mode	Damaging factors
Non-functional barrel	Spring fatigue	– Spring cycling – Stress applied in the leaf
Loss of barrel performance	Wear-and-tear of the leaf spring	– Spring cycling – Lubricant

Table 6.1. *Probable failures of the watch barrel system*

One of the critical points in the design of a barrel spring is the fatigue life of the leaf spring. Indeed, between the fully loaded and fully extended positions, there are strong variations in stresses that can lead to fatigue failure and limited endurance. The accumulation of cycles will damage the part until it breaks.

6.3. Evaluation of the customer usage profile

For this study, we are interested in the usage profiles of customers who invest in a mechanical watch with manual winding.

6.3.1. *Classifying usage typologies*

In order to study the customer watch usage profiles, we evaluated two important quantities: on the one hand, the frequency and time wearing the watch, and, on the other hand, the frequency of winding by customers owning at least one mechanical watch.

We observed a large difference in the use of manual winding. More than half of customers wind their watch more than once a day (see Figure 6.3).

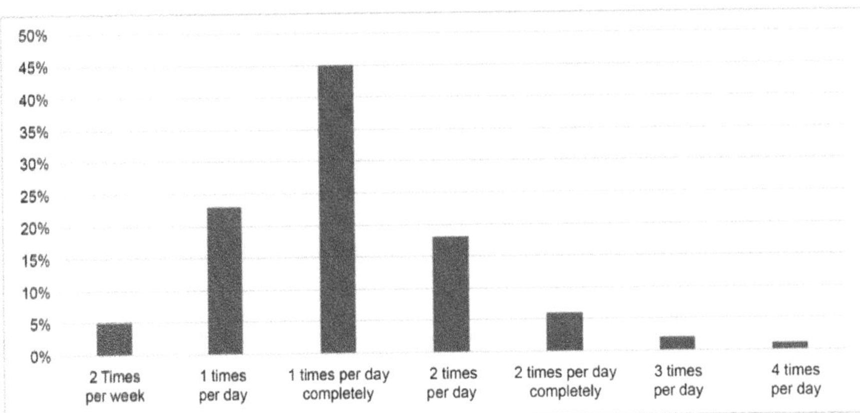

Figure 6.3. *Frequency of customers winding their watches*

In order to quantify barrel usage, we correlated this data with wearing times and wearing frequency. First, it is necessary to cross-reference the data in Figure 6.1 with the frequency of wearing for each customer. Indeed, a customer who winds their watch once a day, but changes their watch every other day, will not have the same influence on the barrel as a customer who winds their watch every day while wearing the same watch 5 days a week.

This correlation between the data allowed us to determine stress profiles by customer.

6.3.2. *Statistical quantification of usage*

From the data processed in the previous section, by associating the winding frequencies with the wearing times and wearing frequencies, we were able to obtain, for each customer, a stress profile. In this study, we are interested in the discharge levels of the barrel. Five levels of discharge amplitude are considered hereafter: 100%, 80%, 60%, 40% and 20%. For each piece of customer data, the number of cycles associated with each step is determined. Figure 6.4 shows two stress profiles from two different customers.

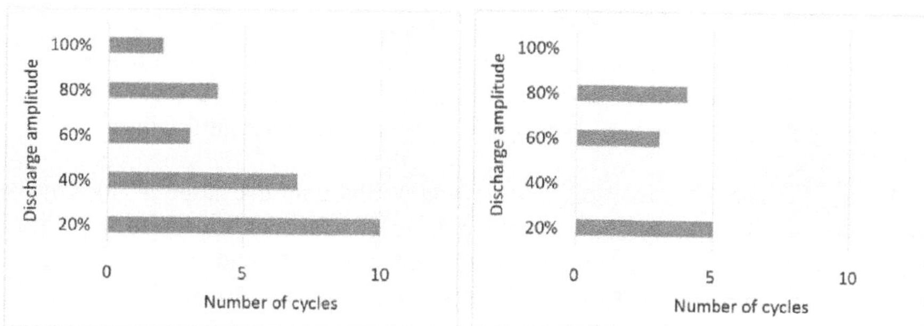

Figure 6.4. *Barrel stress profiles for two different customers – discharge rate versus number of discharge cycles*

Discharge amplitudes in the range of 20–40% correspond to a daily wearing of the watch by the customer and its frequent winding. In this configuration, the barrel is most often fully loaded.

While the amplitudes of 60–100% discharge correspond to less frequent windings (one or two times per week). It is for this last category of wearers that the failure mode studied here is the most stressed. Indeed, it is when the barrel is unloaded to more than 60% that spring fracturing is likely to appear.

These customer usage profiles allow us to determine the law of customer stress associated with the failure mode studied.

6.4. Characterizing experimental reliability

6.4.1. *Performance of failure tests*

In order to better understand the fatigue failure of the barrel spring, we are conducting failure tests for this particular part. The test consists of winding and unwinding the barrel spring within its operating range.

We decide to characterize the life of the springs by loading them according to several levels in order to apply different levels of stress at each test. These levels are defined in relation to the level of loading and unloading of the spring. We will say that a spring is loaded at:

– 100% when it is rolled up in such a way that 100% of its energy is stored, ready to be restored to the gear train;

– 0% when it is unwound to its maximum and can no longer restore any energy and therefore can no longer drive the gear train.

The tests carried out will consist of cycles between the maximum load point of 100% and different discharge points down to 0%. Since these tests must allow us to characterize the life of the barrel springs, we carry out the tests until failure.

During the tests, it turned out that beyond a minimum discharge of 60% of the barrel, that is, a load of 40%, no barrel would break. This confirmed our hypothesis that the blade would break under limited endurance. Beyond a certain level of discharge, the stress applied to the blade is sufficiently low for it to pass into the domain of unlimited endurance [BAT 97].

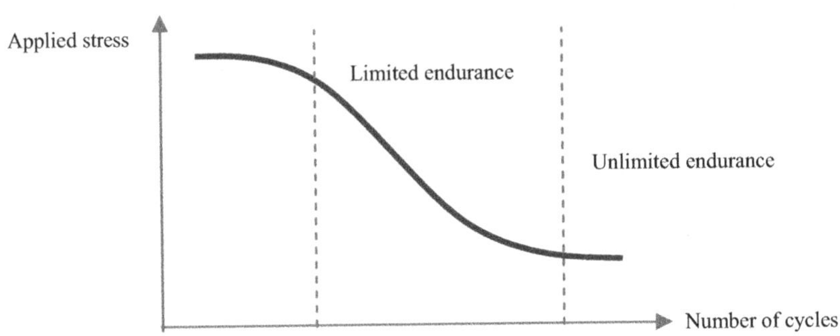

Figure 6.5. *Fatigue curve for the iso-probability of failure*

6.4.2. *Evaluation of the accelerated lifetime law*

We therefore perform failure tests by cycling between the maximum charge point and several discharge levels. The results obtained are shown in Table 6.2.

Amplitude of discharge	Number of cycles to failure
100	560; 469; 417; 544
95	586; 636; 815; 702; 698
90	1,421; 1,369; 710; 1,522; 949
85	2,062; 2,121; 2,380; 2,706; 1,260
75	10,124; 11,543; 14,939; 6,632; 17,897

Table 6.2. *Failure tests cycling between the maximum charge point and several discharge levels*

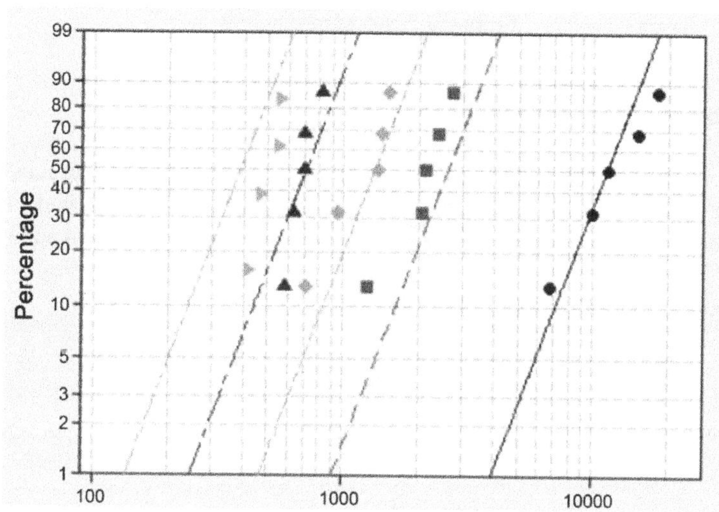

Figure 6.6. *Chart showing failure test cycling between the maximum charge point and several discharge levels*

From these lifespan data, we can reconstruct the fatigue law associated with the studied failure mode:

$$\ln(N) = 59.98 - 11.7 * \ln(amplitude)$$

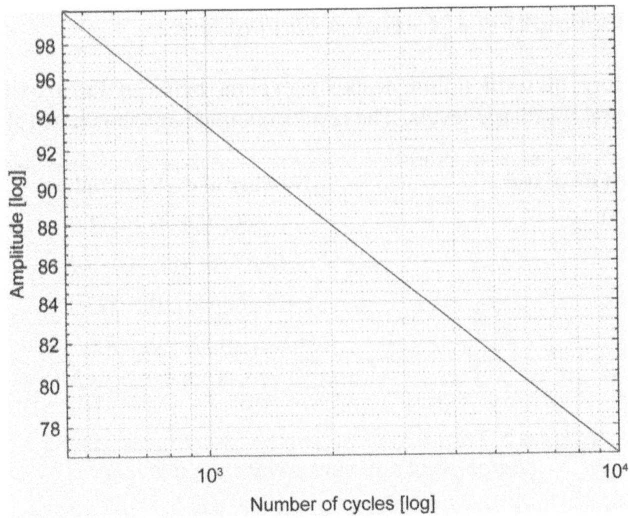

Figure 6.7. *Acceleration law*

6.4.3. *Constructing the law of resistance*

From this data, we can obtain the resistance law of the barrel whatever the test amplitude applied A_{test}. In our case, we place ourselves at a test amplitude 100% to obtain the law of resistance that follows a Weibull law of parameter $\beta = 4$ and $\eta = 426$.

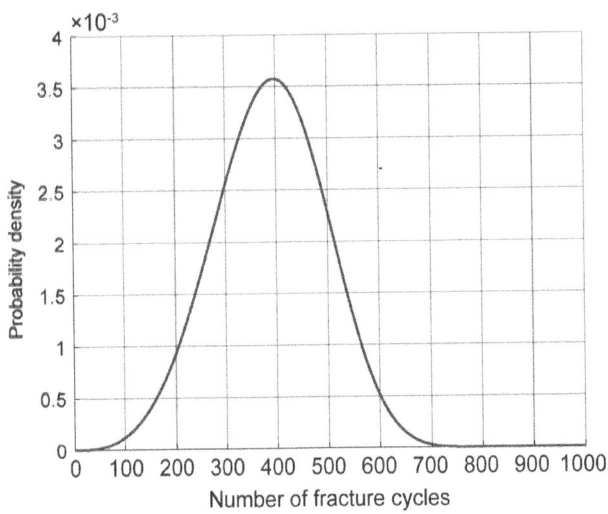

Figure 6.8. *Distribution of the life span at A_{test}*

6.5. Reliability evaluation of customers

Methods borrowed from the automotive industry are transposed to the watchmaking industry and applied to the case study presented in this chapter [PON 04, PON 05a, PON 05b, SOC 17]. These methods have the following objectives:

– evaluate the reliability of customers from test results;

– size a validation plan, that is, determine the number of parts to be tested, the level of stress constraints to be applied, and an acceptance criterion, in order to guarantee a customer reliability objective.

In this application, we are particularly interested in the stress–strength method, well known by reliability specialists. This method is presented in the following section.

6.5.1. *Reliability calculation using the stress–strength method*

Depending on the data from the customer feedback and the possibility to measure the degradation of the studied component or system, several methods can be used to quantify the customer reliability. In the case of the study presented here, the stress–strength method is applied [SOC 17, IMD 06].

This approach is based on the consideration of the variability of the applied stress constraint, random variable denoted as C hereafter (use of watches, environment, etc.) and the variability of the strength resistance of the component or system, a random variable denoted as R (manufacturing dispersion, sizing dispersion, material characteristic dispersion, etc.). The aim of this method is to evaluate the reliability of a component or system, from the distribution of customer stresses and the distribution of strength resistance.

In a first step, a limit state function is defined from these two random variables. This function, denoted G, is defined in R and with values in R, it is described as:

$$G(R, C) = R - C$$

From this function, the failure criterion of the studied component is given as:

$$G(R, C) \leq 0 \Leftrightarrow R \leq C$$

The probability of failure, denoted P_f, is described by:

$$P_f = P(G(R, C) \leq= 0) = P(R \leq C)$$

One of the basic postulates of the stress–strength method is the independence of the stress and strength variables. Let f_R and f_C be the probability densities of strength resistance and stress constraint, respectively. The probability of failure P_f is obtained by:

$$P_f = \int \int f_R(r) f_C(c) dr dc$$

In the case where the random variables R and C are both normal, for parameters (μ_R, σ_R) and (μ_C, σ_C), respectively, this probability is calculated analytically by:

$$P_f = \Phi\left(-\frac{\mu_R - \mu_C}{\sqrt{\sigma_R^2 + \sigma_C^2}} \right)$$

where $\Phi: \mathbb{R} \rightarrow [0,1]$ the distribution function of the reduced centered normal distribution.

In the general case where these two random variables follow arbitrary laws, the use of a statistical estimator is necessary to assess the probability of failure, for which there are many methods that exist in the literature [LEM 14].

The reader interested in a description of the methods for evaluating the probability of failure is invited to refer to [DIV 96, LEM 09, LEM 14, SAP 11].

6.5.2. *Transformation of the stress profile*

The acceleration law, quantified in section 6.3.2, can determine the stress law as applied to customers. This law will be determined for a reference period of 1 year.

For practical reasons, the tests are carried out at a fixed amplitude; it is therefore necessary to transform the customer usage profiles, as mentioned in section 6.3.2, to determine the stress law. We note the A_{test} amplitude level at which the test is successfully performed.

This transformation takes place in several steps. We propose to denote (n_i, a_i) as the coupling (number of cycles, barrel discharge amplitude) for each discharge step $(i \in [\![1,3]\!] : d1 = 60\%, d2 = 80\%, d3 = 100\%)$. According to the previous sections, discharge amplitudes below 60% no longer damage the barrel spring. Consequently, these amplitudes are not taken into account for the next set of calculations.

Using Miner's law of cumulative damage, we can calculate a damage d_j applied by customer j. This damage is written as:

$$dj = \sum_{i=1}^{3} \frac{n_i}{N_i}$$

where N_i are the number of cycles leading to the failure at amplitude a_i. These N_i levels are calculated from the acceleration law determined in section 6.4.2 (see Figure 6.3).

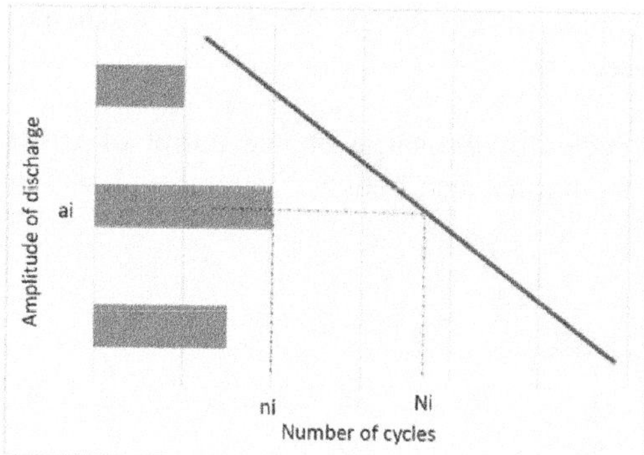

Figure 6.9. *Illustration of the accumulation of discharge amplitudes*

Knowing the cumulative damage caused by customer j, we determine the equivalent number of n_{test} cycles causing damage d_j at amplitude A_{test}. To do this, we use the acceleration law to determine the number of N_{test} cycles that will bring about a failure at amplitude A_{test} (see Figure 6.4).

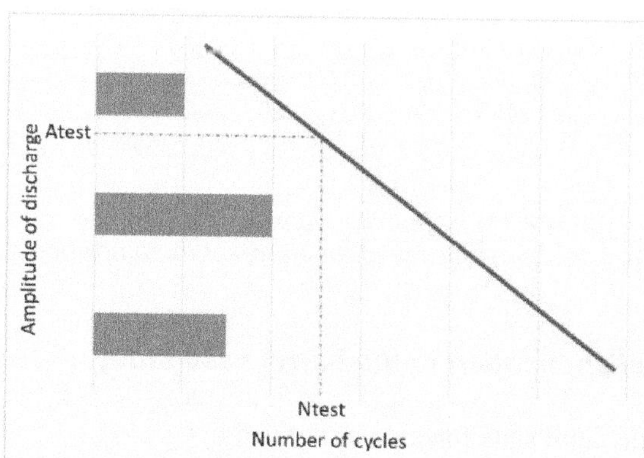

Figure 6.10. *Calculation for the number of N_{test} cycles at amplitude A_{test}*

We can thus write the following relationship:

$$d_j = \frac{n_{test}}{N_{test}}$$

So that:

$$n_{test} = d_j \times N_{test}$$

By repeating this operation for all customers in the database, the statistic for the numbers of cycles applied to customers at the A_{test} discharge amplitude on the barrel spring is derived.

The number of cycles of discharge on the barrel of customers at the A_{test} amplitude follows a log-normal distribution of parameter $\lambda = 4$ and $\beta = 0.7$.

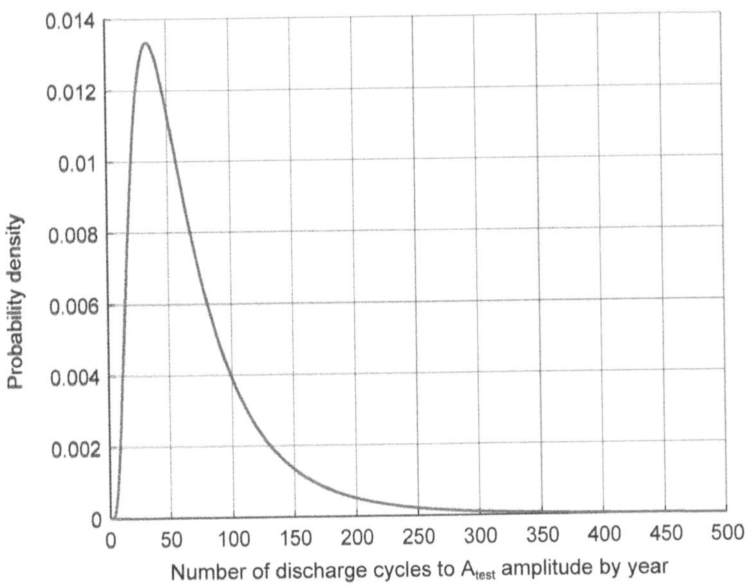

Figure 6.11. *Probability density of the number of annual discharge cycles of the barrel at the amplitude A_{test}*

6.5.3. *Numerical application to the barrel case study*

We consider the limit state function as defined by:

$$G(R, C) = R - C$$

where the resistance strength R is following a Weibull distribution (see section 6.3.3), and the stress constraint C is following a log-normal distribution of parameters (see section 6.4.2).

Figure 6.12 illustrates these two laws.

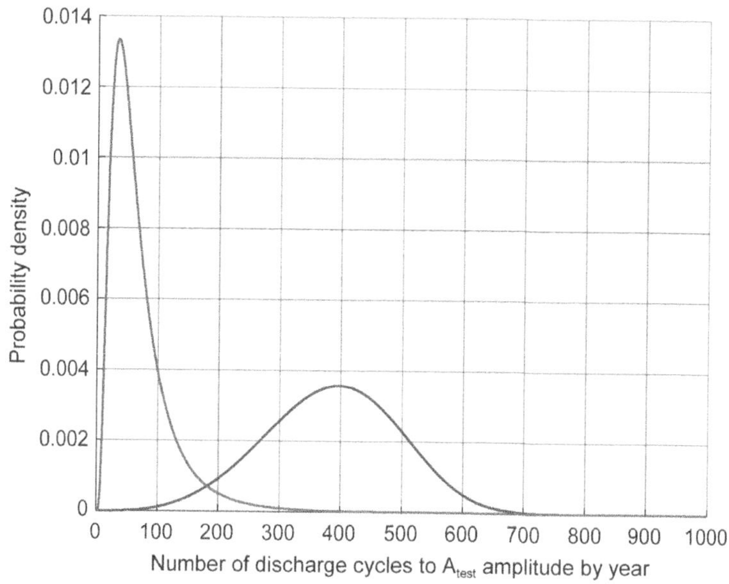

Figure 6.12. *Application of the stress–strength method*

The estimated probability of failure according to the Monte Carlo method is:

$$P_f = 0.81\%$$

with a coefficient variation of 0.5%, which gives this estimate sufficient accuracy.

6.6. Conclusion

Through this example, we wanted to highlight the application of the stress–strength method for the reliability quantification of a watch component.

This technique allows us:

– to highlight the importance of knowing the product's usage profiles to characterize and guarantee the product's reliability over time;

– to combine estimation with accelerated testing so as to reduce and adapt the duration of testing to assist in reducing the time to the market of products.

6.7. References

[BAT 97] BATHIAS C., BAÏLON J.-P., *La fatigue des matériaux et des structures*, Hermes, 1997.

[DIV 96] DITLEVSEN O., MADSEN H.O., *Structural Reliability Methods*, John Wiley and Sons, 1996.

[IMD 06] IMDR–INSTITUT POUR LA MAITRISE DES RISQUES, “Optimisation de la conception et dimensionnement d’un produit à l’aide d’une extension de l’approche résistance/contrainte”, *Projet de l’IMDR-SdF*, no. 1/2005, 2006.

[LEM 09] LEMAIRE M., *Structural Reliability*, ISTE Ltd, London and John Wiley & Sons, New York, 2009.

[LEM 14] LEMAIRE M., *Mechanics and Uncertainty*, ISTE Ltd, London and John Wiley & Sons, New York, 2014.

[PON 05a] PONCET A., SCHIMMERLING P., LYONNET P., et al., “Le dimensionnement des essais de validation par la méthode “contrainte-résistance””, *Congrès Lambda-Mu 12*, 12–14 October 2005.

[PON 05b] PONCET A., Fiabilité expérimentale : dimensionnement optimal d’essais sous contraintes : modélisation des facteurs endommageant le composant et intégration dans le dimensionnement de l’essai. Application à l’automobile, Thesis, Université Jean-Monnet-Saint-Étienne, 2005.

[SAP 11] SAPORTA G., *Probabilités, analyse des données et statistique*, Editions TECHNIP, 2011.

[SOC 17] SOCIÉTÉ DES INGÉNIEURS DE L’AUTOMOBILE, “Guide d’aide à l’estimation et à la validation de la fiabilité automobile”, DC-01-03, July 2017.

7

RBDO of the High Electron Mobility Transistor

In this chapter, we are interested in the study of one of the most important components in complex systems and high-power mechatronic systems, that is, the transistor. The high electron mobility transistor (HEMT) is a technology under development. After having performed a multiphysics modeling of the HEMT, the analysis of its reliability by the two methods SORM and FORM showed the need to optimize the reliability level of the HEMT. In this chapter, we present the reliability-based design optimization (RBDO) applied to HEMT technology in order to improve its performance and reliability. In this sense, the execution of the RBDO requires the development and coupling of two models: the finite element model through Comsol Multiphyisics software, and the RBDO model through Matlab software. The first model is used to simulate the electro-thermomechanical behavior of the component. The second one is used to solve the optimization problem because of its coupling with the first model. After the application of this approach, we succeeded in determining the optimal values of the design variables that allow us to optimize the multiphysical behavior of the structure and the reliability level of the HEMT.

7.1. Introduction

The commissioning of mechatronic systems has been provided with failure and degradation phenomena that are not well controlled, hence the need to identify these failures, in order to ensure a robust and reliable design of these systems [ELH 15a]. Among the very important components in these systems is the transistor. It is an

---

For a color version of all the figures in this chapter, see http://www.iste.co.uk/elhami/applied3.zip.

Chapter written by Abdelhamid AMAR, Bouchaïb RADI and Abdelkhalak EL HAMI.

electronic element that has created a revolution in the design and development of such systems.

The HEMT, based on gallium nitride (GaN), has shown to be efficient in high-power systems. This technology has a very particular structure that works in exceptional conditions, such as high power, high temperature and high frequency applications [ALI 19]. Because of all of these advantages, the HEMT is present in several fields, including telecommunication, electronic warfare and satellites [CHE 19]. Most of the characteristics, such as electron mobility and thermal conductivity, are temperature dependent; these characteristics tend to degrade with the rise of temperature in the structure due to the phenomenon of self-heating. This phenomenon can also give rise to mechanical degradations in the structure, because some physical characteristics of the materials are temperature dependent and deformation dependent at the same time. These degradations can be in the form of displacements, deformations, or stresses. The identification and understanding of these degradations remain an important issue when ensuring the reliability and performance of HEMT technology.

In order to observe the multiphysical behavior of the HEMT, several works have treated the simple modeling of HEMT, either electrically, thermally or electrothermally, through the finite element method [BAC 15, FLO 19], with the aim to study the electrical or thermal behavior of the HEMT, and then identify the interaction between the electrical and thermal parameters during the operation of the component. To this end, they generated simulations to determine the influence of the voltage applied to the transistor terminals on its thermal behavior in general, as well as on the temperature variation in particular [CHA 20]. However, few works have treated the complete modeling (electro-thermomechanical), or even mechanical modeling only, in order to identify the different interactions due to the complexity and number of equations that appear during the complete modeling. Moreover, the realization of an experimental mechanical study of this micro-component is difficult. The modeling allows us to visualize and detect the failures in the structure. On the other hand, the reliability analysis allows us to evaluate the probability of failure according to the variables that cause it [HAM 19a]. There are approaches that can be used to improve the level of reliability [ELH 13]. Reliability optimization allows the reliability of a structure to be improved with a minimum cost. There are several types of optimizations, the classical RBDO has been used in several works, but it shows some disadvantages, either in terms of the stability of the calculation or the weakness of the convergence [ELH 11], because it performs the process of optimization and reliability analysis in an independent way. On the other hand, the hybrid RBDO, according to several works, has shown computational efficiency. It has overcome the shortcomings of the classical approach because the process performs the optimization and reliability analysis simultaneously.

With the aim of studying HEMT technology through a new approach, this chapter has been divided into several sections. First, we will start with a description of the HEMT technology and its uses; next, we will deal with the multiphysics modeling of the component. In the next section, we will define the reliability analysis methods and the different optimization methods that we will use in this study. The last section will present and discuss the results obtained through the application of the hybrid RBDO on the HEMT.

7.2. Description of the HEMT technology

The HEMT, which is based on the composition of aluminum gallium nitride/GaN (AlGaN/GaN), is a very important electronic component due to its functions, structure and materials. GaN is characterized by high mobility, a very high electrical breakdown field and high thermal conductivity. In addition, HEMT is characterized by a special structure consisting of several layers of different materials (Figure 7.1); a silicon carbide layer that represents the substrate on which the component is built, a nucleation layer that serves to achieve mesh tuning between the substrate and GaN, a GaN layer that contains the Gas-2DEG in its upper part and an AlGaN layer that serves to create a heterojunction with the GaN layer and other layers that provide other functions [NIF 17]. After carrying out a literary study, it was found that the HEMT can have several structures. It was also noted that there is a difference between these structures regarding the topology and materials used, all of which leads us to conclude that the high mobility transistor is a technology that is still developing and improving in order to find the best structure of the HEMT to increase its performance and reliability.

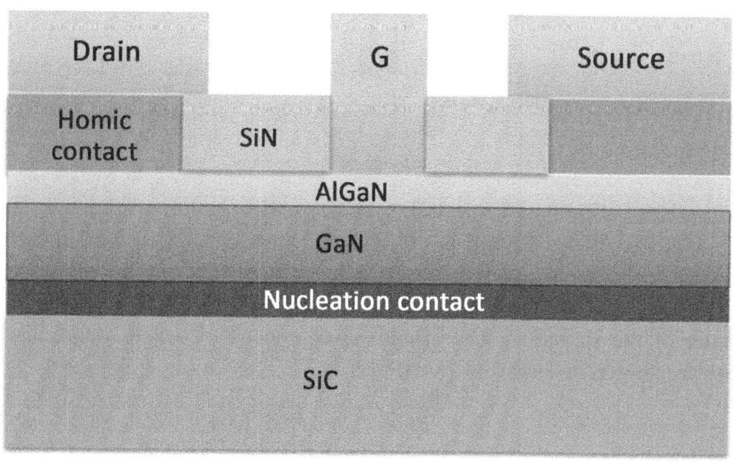

Figure 7.1. *Structure of HEMT*

7.3. Electrothermomechanical modeling of HEMT

In this section, we study the multiphysics behavior of the HEMT in two parts.

7.3.1. *Electrothermal modeling of HEMT*

During its operation, the HEMT dissipates a power at the active zone of the gate output, the dissipated power is a function of the voltage applied to the terminals and the current flowing through the transistor, such that:

$$P_{diss} = V_{ds}.I_{ds} \qquad [7.1]$$

where V_{ds} and I_{ds} are, respectively, the voltage and current that flows between the drain and the source. To define the thermal behavior of the transistor, we must study the heat transfer within the component; this transfer is mainly by conduction, given that convection and radiation are negligible since they represent less than 1.5% [AUB 04]. Based on this mode of heat exchange, the heat equation will be given as [ELH 15a]:

$$k.\nabla^2 T = \rho \cdot C_{P} \cdot \frac{\partial T}{\partial t} + Q \qquad [7.2]$$

where:

– Q: power dissipation in joules;

– K: thermal conductivity ($W \cdot m^{-1} \cdot K^{-1}$);

– ρ: density ($kg \cdot m^{-3}$);

– C_p: thermal mass capacity ($J \cdot kg^{-1} \cdot K^{-1}$);

– T: temperature (K).

This equation allows us to describe the thermal behavior of the transistor. The thermal modeling of our system is done in the form of a 3D model using Comsol Multiphysics software in order to observe the influence of the power dissipated in the active zone of the transistor during its operation, and that of the reference temperature of the substrate. The system is composed of several materials, whose properties are shown in Table 7.1 [MAR 13].

Materials	Volume density, ρ (kg/m^3)	Thermal conductivity, K (W/m/K)	Thermal mass capacity, Cp (J/kg/K)
Au	19,300	310	137
SiN	3,300	10	713
AlGaN	5,470	$25\cdot\left(\frac{293}{273+T}\right)^{-1.35}$	548
GaN	6,100	$161\cdot\left(\frac{293}{273+T}\right)^{-1.45}$	490
SiC	3,220	$416\cdot\left(\frac{293}{273+T}\right)^{-1.5}$	690

Table 7.1. *Thermal properties of materials*

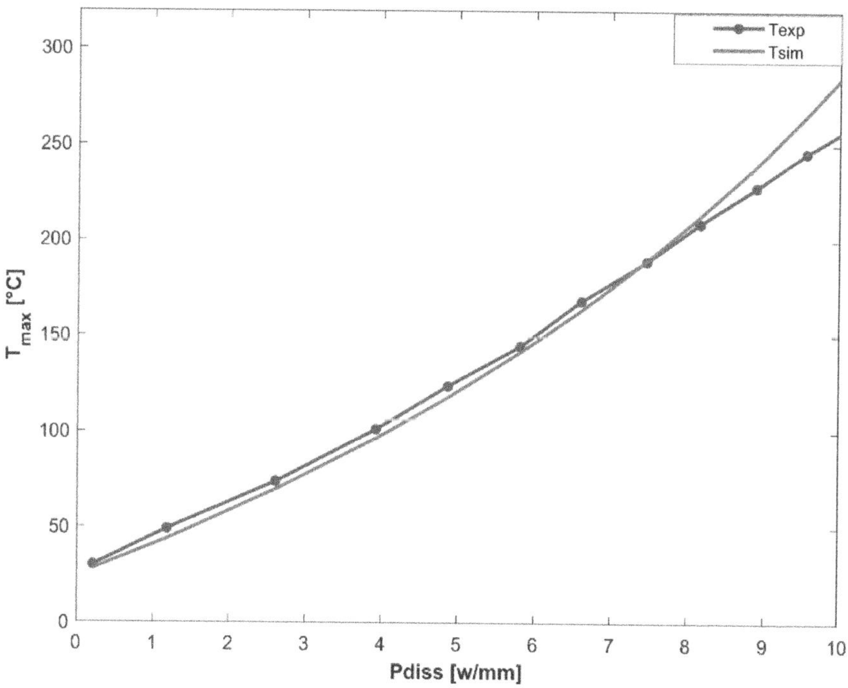

Figure 7.2. *Evolution of the temperature as a function of the dissipated power*

Using these properties, the finite element model was developed using Comsol Multiphysics software. After the electrothermal model was completed, the results of the numerical simulations are presented in Figure 7.2, and a comparison with the experimental results is presented. From this figure, it can be seen that the temperature of the HEMT operation increases as a function of the dissipated power following a nonlinear evolution. We can also notice a good correlation between the experimental and simulation results, which allows us to validate our model.

7.3.2. *Thermomechanical modeling of HEMT*

In this part, we will start the thermomechanical modeling, which allows us to study the influence of the thermal behavior on the mechanical behavior of the HEMT structure. The distribution and increase in the temperature within the HEMT, due to the dissipated power, can give rise to phenomena and mechanical degradations in the HEMT structure. In order to identify these phenomena, we will develop a thermomechanical model. This model makes it possible to link the thermal parameters, such as temperature, with the mechanical parameters, such as stress constraints, deformations and displacements. The coupling of the thermal model to the mechanical model can be translated into equations. Because of the equations for thermoelasticity, we can define the deformations and entropy density by the following expressions [ELH 15a]:

$$\{\varepsilon\} = [D]^{-1}\{\sigma\} + \{\alpha\}\Delta T \qquad [7.3]$$

$$\{S\} = \{\alpha\}^T\{\sigma\} + \frac{\rho \cdot C_P}{T_0}\Delta T; \Delta T = T - T_{\text{ref}} \qquad [7.4]$$

where $\{\varepsilon\}$ is the total deformation vector; it is a function of stress vector, thermal expansion coefficient vector, temperature difference and inverse of the elastic stiffness matrix.

In equation [7.4], the entropy is a function of several thermal and mechanical parameters, such as stresses and temperature variations, among others.

According to the law of thermodynamics, entropy density can be expressed as a function of heat density, according to the following equation:

$$Q = T_0 \cdot S \qquad [7.5]$$

On the other hand, the vector of thermoelastic coefficients is expressed as follows:

$$\{\beta\} = [D]\{\alpha\} \tag{7.6}$$

Noting also the specific heat at constant volume C_v, its expression is given as:

$$C_v = C_p - \frac{T_0}{T}\{\alpha\}^T\{\beta\} \tag{7.7}$$

The integration between the previous equations allows us to obtain new formulas for the principal equations, which thus become [ELH 15a]:

$$\{\varepsilon\} = [D]\{\varepsilon\} - \{\beta\}\Delta T \tag{7.8}$$

$$Q = T_0\{\beta\}^T\{\varepsilon\} + \rho \cdot C_v \cdot \Delta T \tag{7.9}$$

These two equations ensure the thermomechanical coupling, linking the thermal parameters with the mechanical parameters. The two elementary couplings, electrothermal and thermomechanical, ensure the global electro-thermomechanical coupling. The physical properties of the materials needed to realize the model are listed in Table 7.2 [STE 13]. Some properties are already integrated in the Comsol Multiphysics material library.

Material	E (GPa)	CTE (1/K)	U
Au	70	14.2	0.44
SiN	250	1.5e-6	0.27
AlGaN	212.75	5.2425e-6	0.319
GaN	181	8.6e-6	0.352
SiC	748	4.3e-6	0.45

Table 7.2. *Physical properties of the materials*

The finite element model that we have developed under Comsol Multiphysics in three dimension has allowed us to observe the distribution of von Mises stresses within the HEMT structure. From Figure 7.3, we can see that the stresses are very important at the level of the substrate, which represents the construction base of the component, and both sides of the grid. These stresses anticipate a value of 4.2 GPa for the critical zones of the structure. The operating conditions of the component have a great mechanical impact on the structure of the HEMT.

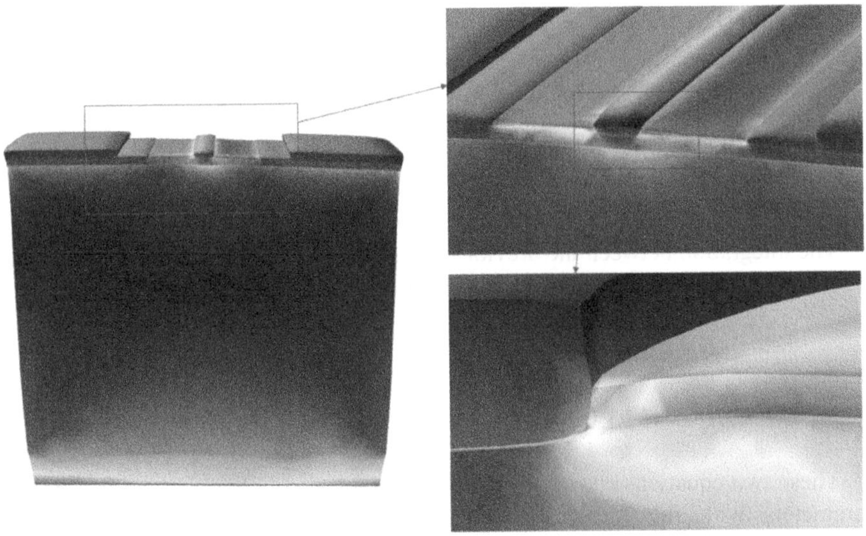

Figure 7.3. *Distribution of von Mises stresses in the HEMT structure*

7.4. Reliability methods

Reliability can be defined in different ways. Moreover, the notion of reliability is linked to several statistical and probabilistic parameters and also to approximation methods described in Chapter 4.

7.5. Reliability analysis of HEMT

Thermal modeling has identified the parameters that have a significant impact on operating temperature, such as the power dissipation P_{diss}, the reference temperature T_{ref} and also the material properties. To study the reliability of the HEMT, it is necessary to construct the performance function, in our case, to ensure the component reliability if the maximum displacement does not exceed $D_{max} = 0.1$ µm, for which the limit state function is expressed as follows:

$$G(X) = D_{max} - D(X) \tag{7.10}$$

D(X) is obtained from the previously developed finite element model, as a function of the vector of random variables X, which represents the parameters of the layers that have a significant effect on the electro-thermomechanical behavior of the HEMT; the relevant layers are the substrate, GaN and AlGaN. In this study, the material properties are considered as random variables with a normal distribution.

Variable	Distribution	Average	Standard deviation
E_{SiC}	Normal	748	74.8
E_{GaN}	Normal	181	18.1
E_{AlGaN}	Normal	212.75	21.275
CTE_{SiC}	Normal	4.3 e-6	4.3 e-7
CTE_{GaN}	Normal	8.6 e-6	8.6 e-7
CTE_{AlGaN}	Normal	5.242 e-6	5.242 e-7
v_{Sic}	Normal	0.45	0.03
v_{GaN}	Normal	0.352	0.03
v_{AlGaN}	Normal	0.319	0.03

Table 7.3. *Distribution of the parameters*

The reliability analysis for HEMT is done by coupling the finite element method, using the Comsol Multiphysics software, and the probabilistic model, using Matlab software. The results of the reliability analysis of HEMT by both SORM and FORM methods are classified in Table 7.4. The level of reliability by the FORM method is 0.83, and by the SORM method it is 0.98; the SORM method provides the more accurate result. From this part of the reliability analysis, it can be seen that the reliability level of the HEMT should be optimized to increase its performance.

	FORM	SORM
β	0.83	0.98
P_f	20.28	16.12
R	79.72	83.87

Table 7.4. *Results of the HEMT reliability analysis*

7.6. Reliability optimization of systems

Optimization is a process that allows us to improve an already designed system or determine the parameters that allow for the design a new system, with minimal cost and high quality. There are several types of optimizations. Among these is the deterministic optimization [HAM 19b], which serves to minimize the objective function under the constraint of deterministic parameters only. This method is expressed as follows:

$$\begin{array}{ll}\min_x & \left\{\begin{array}{l} f(x) \\ g_k(x) \leq 0, k = 1,\ldots,K \end{array}\right. \\ \text{s.t.} & \end{array} \qquad [7.11]$$

where x represents the vector of deterministic variables and $g_k(x)$ is the geometric and physical feasibility function; this represents the upper and lower limits of optimization vector variables x.

However, in order to control the level of performance, it is necessary to integrate the reliability analysis into the optimization process. In this sense, another form of optimization is the reliability optimization: RBDO is very effective in the field of structures and allows us to determine the best possible design in terms of cost and quality; moreover, it takes into consideration the level of reliability. RBDO is based on failure analysis to measure the effect of uncertainties in the system. Currently, it is considered a popular philosophy for solving problems across diverse fields. There are several forms of RBDO that will be discussed throughout the rest of this chapter.

7.6.1. *The classic RBDO approach*

The classical RBDO approach is based on solving the reliability problem and the optimization problem independently. After performing the reliability analysis, it sends the results back to the optimization process to be considered as constraints in order to minimize the objective function. The classical RBDO method is formed by two subproblems that run in two different spaces: the physical space and the normalized space. In order to minimize the objective function f(x), the classical RBDO is expressed as follows:

$$\begin{array}{ll}\min_x & \left\{\begin{array}{l} f(x) \\ g_k(x) \leq 0, k = 1,\ldots,K \\ \beta(x,u) \geq \beta_t \end{array}\right. \\ \text{s.t.} & \end{array} \qquad [7.12]$$

where u represents the vector modulus in the normalized space, $g_k(x)$ represents the deterministic constraints, β_t the target reliability index and $\beta(x,u)$ the system reliability index as a function of deterministic and random variables. The reliability analysis is done by calculating the reliability index that represents the minimum distance between the origin and the limit state function $H(x,u)$ in the normalized space. The calculation is done by solving the following optimization problem:

$$\begin{array}{ll} \min_x & \left\{ \begin{array}{l} d(u) = \sqrt{\sum u_i^2} \\ H(x,u) \leq 0 \end{array} \right. \\ \text{s.t.} & \end{array} \qquad [7.13]$$

The classical RBDO approach alternates between reliability calculation and objective function minimization. As a result, it has low computational efficiency and convergence stability [ELH 08].

7.6.2. *The hybrid RBDO approach*

In contrast to the classical approach, the hybrid RBDO approach allows us to simultaneously solve the reliability analysis problem and the optimization problem in a new space called the hybrid space. In this approach, the objective function takes a new form that links the objective function and the reliability level. Therefore, the formulation of the hybrid RBDO becomes [ELH 08]:

$$\min_{x,y} F(x,y) = f(x) \cdot d_\beta(x,y)$$
$$s.t \left\{ \begin{array}{l} G(x,y) \leq 0 \\ g_k(x,y) \leq 0, k = 1, \ldots, K \\ d_\beta(x,y) \geq \beta_t \end{array} \right. \qquad [7.14]$$

The goal is to minimize the new objective function $\mathrm{F(x,y)}$, which is a function of the standard objective function $\mathrm{f(x)}$ and the distance $d_\beta(x,y)$ between the optimal point x and the design point y in the hybrid space, under the constraints of the limit state function $\mathrm{G(x,y)}$, deterministic constraints $g_k(x)$ and the target reliability index β_t. In this way, the hybrid method overcomes the drawbacks of the classical approach.

7.7. HEMT reliability optimization using the hybrid RBDO approach

7.7.1. Description of the optimization problem

During HEMT operations, the interaction between the electrical and thermal phenomena can generate mechanical phenomena in the structure of the component, in the form of displacements, deformations and stresses. These can lead to the failure of the transistor. Therefore, the control and minimization of the effect of these phenomena represent a challenge that seriously influences the reliability of the component. In this sense, the aim of our study is to minimize the effect of Von mises stresses on the HEMT structure under deterministic and random constraints, without neglecting the reliability level. In the previous section, two RBDO methods were mentioned, while the suitable method in our case is the hybrid method, because it takes into consideration the reliability level during the optimization process. Therefore, the problem of optimization by the hybrid approach is expressed as follows:

$$\min_{x,y} F(x,y) = \sigma_e(x) \cdot d_\beta(x,y)$$
$$s.t \begin{cases} d_\beta(x,y) \geq \beta_t \\ g_k(x,y) \leq 0, \\ k = 1,\ldots,K \end{cases} \qquad [7.15]$$

where x and y are, respectively, the deterministic and random variables in the hybrid space, $\sigma_e(x)$ are the von Mises stresses at the structural level, $d_\beta(x,y)$ is the distance between the design point and the optimal point, $F(x,y)$ is the objective function to be minimized, β_t is the desired reliability index and $g_k(x,y)$ represents the deterministic and random constraints.

7.7.2. Results and discussion

In this section, we present the results of the optimization process that we have developed previously. The execution of this process requires a coupling between two models: the finite element method, which we have performed using the Comsol Multiphysics software, and the optimization-reliability model, which we have performed using the Matlab software. The three geometrical parameters, (a) substrate thickness, (b) GaN thickness, and (c) AlGaN thickness, were chosen as the design variables since they have a great influence on the electro-thermomechanical behavior of the HEMT. Next, the initial, minimum and maximum values needed to solve this problem are classified in Table 7.5.

Parameters (μm)	Deterministic	Minimums	Maximums
A	100	90	160
B	1.7	1.1	1.9
C	0.03	0.01	0.09

Table 7.5. *Results of the reliability analysis of the HEMT transistor*

The results obtained by the hybrid RBDO approach are mentioned in Table 7.6. This table shows the initial and optimal values of the variables; it was also found that the hybrid RBDO method minimized the effect of the Von stresses placed on the HEMT structure. After 30 iterations, the value of the stresses decreased from 4.2 to 3.016 GPa and the reliability level improved from 0.83 to 3.08. It can also be noticed that the values for the optimal variables are different from the initial values.

Parameters		Initial Values x0	Initial Values y0	Design Point y	Optimal Solution x
Design variables	a	100	105.1	159.95	98.4164
	b	1.7	1.65	1.5466	1.4293
	c	0,03	0.042	0.0584	0.0545
Objective function		4.2e+09	–	–	3.0163e+09
Level of reliability		0.83	–	–	3.08

Table 7.6. *Results of the reliability analysis of the HEMT transistor*

7.8. Conclusion

In this chapter, an approach to improve HEMT reliability has been proposed using the hybrid RBDO method. First, the HEMT reliability level was evaluated by the FORM and SORM approximation methods. In order to optimize the reliability level, the process of the hybrid RBDO method is used by coupling two models. The first model is the finite element model, developed using Comsol Multiphysics software, which allows us to simulate the multiphysics behavior of the HEMT and to observe the thermomechanical phenomena of its structure. The second model is the RBDO probabilistic model, coded using Matlab software, which is used to simultaneously minimize the objective function under stress constraints and evaluate the level of reliability. The application of the RBDO approach on the HEMT technology allowed us to define new design parameters. The obtained results allowed for the minimization of von Mises stresses on the structure and improved its reliability level. The hybrid RBDO method has shown an efficiency to solve optimization problems when compared against the classical method and the deterministic method, despite the fact that this approach remains costly in terms of both time and the stability of the calculation.

7.9. References

[ABO 11a] ABO AL-KHEER A., EID M., AOUES Y. et al., "Theoretical analysis of the spatial variability in tillage forces for fatigue analysis of tillage machines", *Journal of Terramechanics*, vol. 48, no. 4, pp. 285–295, 2011.

[ABO 11b] ABO AL-KHEER A., KHARMANDA M.G., EL HAMI A. et al., "Estimating the variability of tillage forces on a chisel plough shank by modeling the variability of tillage system parameters", *Computers and Electronics in Agriculture*, vol. 78, no. 1, pp. 61–70, 2011.

[ABO 11c] ABO AL-KHEER A., EL-HAMI A., KHARMANDA M.G. et al., "Reliability-based design for soil tillage machines", *Journal of Terramechanics*, vol. 48, no. 1, pp. 57–64, 2011.

[ALI 19] ALIM M.A., ALI M.M., REZAZADEH A.A. et al., "Thermal response for intermodulation distortion components of GaN HEMT for low and high frequency applications", *Microelectronic Engineering*, vol. 209, pp. 53–59, 2019.

[AUB 04] AUBRY, R., Etude des aspects électrothermiques de la filière HEMT AlGaN/GaN pour application de puissance hyperfréquence, PhD Thesis, Université des sciences et technologiques de Lille, 2004.

[BAC 15] BACZKOWSKI L., JACQUET J.C., JARDEL O. et al., "Temperature measurements in RF operating conditions of AlGaN/GaN HEMTs using IR microscopy and Raman spectroscopy", *European Microwave Week 2015*, vol. 5, pp. 152–155, 2015.

[BEN 13] BEN ABDESSALEM A., PAGNACCO E., EL-HAMI A., "Increasing the stability of T-shape tube hydroforming process under stochastic framework", *International Journal of Advanced Manufacturing Technology*, vol. 69, nos. 5–8, pp. 1343–1357, 2013.

[BEN 15] BEN ABDESSALEM A., EL HAMI A., "A probabilistic approach for optimising hydroformed structures using local surrogate models to control failures", *International Journal of Mechanical Sciences*, vol. 96–97, pp. 143–162, 2015.

[BOU 96] BOUYEKHF R., EL MOUDNI A., EL HAMI A. et al., "Reduced order modelling of two-time-scale discrete non-linear systems", *Journal of the Franklin Institute*, vol. 333, no. 4, pp. 499–512, 1996.

[BOU 01] BOUYEKHF R., EL HAMI A., EL MOUDNI A., "Optimal control of a particular class of singularly perturbed nonlinear discrete-time systems", *IEEE Transactions on Automatic Control*, vol. 46, no. 7, pp. 1097–1101, 2001.

[CHA 20] CHATTERJEE B., DUNDAR C., BEECHEM T.E. et al., "Nanoscale electro-thermal interactions in AlGaN/GaN high electron mobility transistors", *Journal of Applied Physics*, vol. 127, no 4, p. 044502, 2020.

[CHE 08] CHEROUAT A., RADI B., EL HAMI A., "The frictional contact of the composite fabric's shaping", *Acta Mechanica*, vol. 199, nos. 1–4, pp. 29–41, 2008.

[CHE 19] CHEN Y., XU Y., WANG F. et al., "Improved quasi-physical zone division model with analytical electrothermal Ids model for AlGaN/GaN heterojunction high electron mobility transistors", *International Journal of Numerical Modelling: Electronic Networks, Devices and Fields*, vol. 33, pp. 1–17, 2019.

[DAM 17] DAMMAK K., EL HAMI A., KOUBAA S. et al., "Reliability based design optimization of coupled acoustic-structure system using generalized polynomial chaos", *International Journal of Mechanical Sciences*, vol. 134, pp. 75–84, 2017.

[DAM 19] DAMMAK K., KOUBAA S., EL HAMI A. et al., "Numerical modelling of vibro-acoustic problem in presence of uncertainty: Application to a vehicle cabin", *Applied Acoustics*, vol. 144, pp. 113–123, 2019.

[ELH 93] EL HAMI A., LALLEMENT G., MINOTTI P. et al., "Methods that combine finite group theory with component mode synthesis in the analysis of repetitive structures", *Computers and Structures*, vol. 48, no. 6, pp. 975–982, 1993.

[ELH 96] EL HAMI A., RADI B., "Some decomposition methods in the analysis of repetitive structures", *Computers and Structures*, vol. 58, no. 5, pp. 973–980, 1996.

[ELH 08] EL HAMI A., RADI B., "Analysis of the dynamic contact inside the piezoelectric motor using RBDO", *Applied Mathematical Sciences*, vol. 2, no. 42, pp. 2047–2062, 2008.

[ELH 09] EL HAMI A., RADI B., CHEROUAT A., "Treatment of the composite fabric's shaping using a Lagrangian formulation", *Mathematical and Computer Modelling*, vol. 49, nos. 7–8, pp. 1337–1349, 2009.

[ELH 11] EL HAMI A., RADI B., "Comparison study of different reliability-based design optimization approaches", *Advanced Materials Research*, vol. 274, pp. 113–121, 2011.

[ELH 13] EL HAMI A., RADI, B., *Uncertainty and Optimization in Structural Mechanics*, ISTE Ltd, London and John Wiley & Sons, New York, 2013.

[ELH 15a] EL HAMI A., POUGNET P., *Embedded Mechatronic Systems 1: Analysis of Failures, Predictive Reliability*, ISTE Press Ltd, London and Elsevier Ltd, Oxford, 2015.

[ELH 15b] EL HAMI A., POUGNET P., *Embedded Mechatronic Systems 2: Analysis of Failures, Modeling, Simulation and Optimization*, ISTE Press Ltd, London and Elsevier Ltd, Oxford, 2015.

[FLO 19] FLOROVIČ M., SZOBOLOVSZKÝ R., KOVÁČ J. et al., "Rigorous channel temperature analysis verified for InAlN/AlN/GaN HEMT", *Semiconductor Science and Technology*, vol. 34, no. 6, p. 065021, 2019.

[GUE 15a] GUERINE A., EL HAMI A., WALHA L. et al., "Approach for the dynamic analysis of one stage gear system with uncertain parameters", *Mechanism and Machine Theory*, vol. 92, pp. 113–126, 2015.

[GUE 15b] GUERINE A., EL HAMI A., FAKHFAKH T. et al., "A polynomial chaos method to the analysis of the dynamic behavior of spur gear system", *Structural Engineering and Mechanics*, vol. 53, no. 4, pp. 819–831, 2015.

[GUE 16a] GUERINE A., EL HAMI A., WALHA L. et al., "A polynomial chaos method for the analysis of the dynamic behavior of uncertain gear friction system", *European Journal of Mechanics, A/Solids*, vol. 59, pp. 76–84, 2016.

[GUE 16b] GUERINE A., EL HAMI A., WALHA L. et al., "Dynamic response of a Spur gear system with uncertain friction coefficient", *Advances in Engineering Software*, vol. 120, pp. 45–54, 2016.

[GUE 17] GUERINE A., EL HAMI A., WALHA L. et al., "Dynamic response of wind turbine gear system with uncertain-but-bounded parameters using interval analysis method", *Renewable Energy*, vol. 113, pp. 679–687, 2017.

[HAM 19a] HAMDANI H., EL HAMI A., RADI B., "Reliability analysis of tape based chip-scale packages based metamodel", *Microelectronics Reliability*, vol. 102, p. 113445, 2019.

[HAM 19b] HAMDANI H., RADI B., EL HAMI A., "Metamodel assisted evolution strategies for global optimization of solder joints reliability in embedded mechatronic devices", *Microsystem Technologies*, vol. 25, no. 10, pp. 3801–3812, 2019.

[HUA 16] HUANG C., RADI B., EL HAMI A., "Uncertainty analysis of deep drawing using surrogate model based probabilistic method", *Manufacturing Technology*, vol. 86, nos. 9–12, pp. 3229–3240, 2016.

[JAN 17] JANNOUN M., AOUES Y., PAGNACCO E. et al., "Probabilistic fatigue damage estimation of embedded electronic solder joints under random vibration", *Microelectronics Reliability*, vol. 78, pp. 249–257, 2017.

[KHA 14] KHARMANDA G., IBRAHIM M.-H., ABO AL-KHEER A. et al., "Reliability-based design optimization of shank chisel plough using optimum safety factor strategy", *Computers and Electronics in Agriculture*, vol. 109, pp. 162–171, 2014.

[MAB 17] MABROUK I.B., EL HAMI A., WALHA L. et al., "Dynamic vibrations in wind energy systems: application to vertical axis wind turbine", *Mechanical Systems and Signal Processing*, vol. 85, pp. 396–414, 2017.

[MAK 16] MAKHLOUFI A., AOUES Y., EL HAMI A., "Reliability based design optimization of wire bonding in power microelectronic devices", *Microsystem Technologies*, vol. 22, no. 12, pp. 2737–2748, 2016.

[MAR 13] MARCON D., MENEGHESSO G., WU T. et al., Reliability analysis of permanent degradations, *IEEE Transactions on Electron Devices*, vol. 60, no. 10, pp. 3132–3141, 2013.

[MOR 02] MORO T., EL HAMI A., EL MOUDNI A., "Reliability analysis of a mechanical contact between deformable solids", *Probabilistic Engineering Mechanics*, vol. 17, no. 3, pp. 227–232, 2002.

[NIF 17] NIFA I., LEROUX C., TORRES A. et al., "Characterization of 2DEG in AlGaN/GaN heterostructure by Hall effect", *Microelectronic Engineering*, vol. 178, pp. 128–131, 2017.

[RAD 07] RADI B., EL HAMI A., "Reliability analysis of the metal forming process", *Mathematical and Computer Modelling*, vol. 45, nos. 3–4, pp. 431–439, 2007.

[SAR 11] SARSRI D., AZRAR L., JEBBOURI A. et al., "Component mode synthesis and polynomial chaos expansions for stochastic frequency functions of large linear FE models", *Computers and Structures*, vol. 89, nos. 3–4, pp. 346–356, 2011.

[STE 13] STEVENS L.E., Thermo-Piezo-Electro-Mechanical Simulation of AlGaN (Aluminum Gallium Nitride)/GaN (Gallium Nitride) High Electron Mobility Transistor, Thesis, Utah State University, 2013.

List of Authors

Abdelhamid AMAR
INSA-Rouen-Normandie
France

David DELAUX
Valeo
La Verrière
France

Abdelkhalak EL HAMI
INSA-Rouen-Normandie
France

Philippe EUDELINE
Normandie AeroEspace (NAE)
France

Henri GRZESKOWIAK
IIGC
Versailles
France

Eric MANOUVRIER
Valeo
Etaples
France

Medoune NDIAYE
Stellantis
Vélizy
France

Ludovic NGAVOUKA
Valeo
Creteill
France

Anthony PONCET
Richemont
Buttes
Switzerland

Bouchaïb RADI
FST-Settat
Morocco

Caroline RAMUS-SERMENT
Stellantis
Sochaux-Belchamp
France

Matthieu SALLIN
Richemont
Buttes
Switzerland

Claire SCHAYES
Valeo
Lille
France

Paul SCHIMMERLING
Retired Expert
Paris
France

Index

W, Z

Summary of Volume 1

Chapter 6. CMA-ES Assisted by the Kriging Metamodel for the Optimization of Thermomechanical Performances of Mechatronic Packaging

Hamid HAMDANI, Bouchaib RADI and Abdelkhalak EL HAMI

Chapter 7. Reliable Optimization of Vibro-acoustic Problems in the Presence of Uncertainties via Polynomial Chaos

Khalil DAMMAK and Abdelkhalak EL HAMI

Summary of Volume 2

Chapter 2. Fatigue Damage Analysis and Reliability Optimization of Structures Subjected to Random Vibrations
Ahmed YAICH and Abdelkhalak EL HAMI

Chapter 3. Accelerated Testing
Henri GRZESKOWIAK, David DELAUX and Abdelkhalak EL HAMI

Other titles from

iSTE

in

Mechanical Engineering and Solid Mechanics

2022

BAYLE Franck
Product Maturity 1: Theoretical Principles and Industrial Applications
(Reliability of Multiphysical Systems Set – Volume 12)
Product Maturity 2: Principles and Illustrations
(Reliability of Multiphysical Systems Set – Volume 13)

EL HAMI Abdelkhalak, DELAUX David, GRZESKOWIAK Henri
Reliability and Physics-of-Healthy in Mechatronics
(Reliability of Multiphysical Systems Set – Volume 15)

LANNOY André
Reliability of Nuclear Power Plants: Methods, Data and Applications
(Reliability of Multiphysical Systems Set – Volume 14)

LEDOUX Michel, EL HAMI Abdelkhalak
Heat Transfer 3: Convection, Fundamentals and Monophasic Flows
(Mathematical and Mechanical Engineering Set – Volume 11)
Heat Transfer 4: Convection, Two-Phase Flows and Special Problems
(Mathematical and Mechanical Engineering Set – Volume 12)

PLANCHETTE Guy
Cindynics, The Science of Danger: A Wake-up Call
(Reliability of Multiphysical Systems Set – Volume 11)

2021

CHALLAMEL Noël, KAPLUNOV Julius, TAKEWAKI Izuru
Modern Trends in Structural and Solid Mechanics 1: Static and Stability
Modern Trends in Structural and Solid Mechanics 2: Vibrations
Modern Trends in Structural and Solid Mechanics 3: Non-deterministic Mechanics

DAHOO Pierre Richard, POUGNET Philippe, EL HAMI Abdelkhalak
Applications and Metrology at Nanometer Scale 1: Smart Materials, Electromagnetic Waves and Uncertainties
(Reliability of Multiphysical Systems Set – Volume 9)
Applications and Metrology at Nanometer Scale 2: Measurement Systems, Quantum Engineering and RBDO Method
(Reliability of Multiphysical Systems Set – Volume 10)

LEDOUX Michel, EL HAMI Abdelkhalak
Heat Transfer 1: Conduction
(Mathematical and Mechanical Engineering Set – Volume 9)
Heat Transfer 2: Radiative Transfer
(Mathematical and Mechanical Engineering Set – Volume 10)

2020

SALENÇON Jean
Elastoplastic Modeling

2019

BAYLE Franck
Reliability of Maintained Systems Subjected to Wear Failure Mechanisms: Theory and Applications
(Reliability of Multiphysical Systems Set – Volume 8)

BEN KAHLA Rabeb, BARKAOUI Abdelwahed, MERZOUKI Tarek
Finite Element Method and Medical Imaging Techniques in Bone Biomechanics
(Mathematical and Mechanical Engineering Set – Volume 8)

IONESCU Ioan R., QUEYREAU Sylvain, PICU Catalin R., SALMAN Oguz Umut
Mechanics and Physics of Solids at Micro- and Nano-Scales

LE VAN Anh, BOUZIDI Rabah
Lagrangian Mechanics: An Advanced Analytical Approach

MICHELITSCH Thomas, PÉREZ RIASCOS Alejandro, COLLET Bernard, NOWAKOWSKI Andrzej, NICOLLEAU Franck
Fractional Dynamics on Networks and Lattices

SALENÇON Jean
Viscoelastic Modeling for Structural Analysis

VÉNIZÉLOS Georges, EL HAMI Abdelkhalak
Movement Equations 5: Dynamics of a Set of Solids
(Non-deformable Solid Mechanics Set – Volume 5)

2018

BOREL Michel, VÉNIZÉLOS Georges
Movement Equations 4: Equilibriums and Small Movements
(Non-deformable Solid Mechanics Set – Volume 4)

FROSSARD Etienne
Granular Geomaterials Dissipative Mechanics: Theory and Applications in Civil Engineering

RADI Bouchaib, EL HAMI Abdelkhalak
Advanced Numerical Methods with Matlab® 1: Function Approximation and System Resolution
(Mathematical and Mechanical Engineering Set – Volume 6)
Advanced Numerical Methods with Matlab® 2: Resolution of Nonlinear, Differential and Partial Differential Equations
(Mathematical and Mechanical Engineering Set – Volume 7)

SALENÇON Jean
Virtual Work Approach to Mechanical Modeling

2017

BOREL Michel, VÉNIZÉLOS Georges
Movement Equations 2: Mathematical and Methodological Supplements (Non-deformable Solid Mechanics Set – Volume 2)
Movement Equations 3: Dynamics and Fundamental Principle (Non-deformable Solid Mechanics Set – Volume 3)

BOUVET Christophe
Mechanics of Aeronautical Solids, Materials and Structures
Mechanics of Aeronautical Composite Materials

BRANCHERIE Delphine, FEISSEL Pierre, BOUVIER Salima, IBRAHIMBEGOVIC Adnan
From Microstructure Investigations to Multiscale Modeling: Bridging the Gap

CHEBEL-MORELLO Brigitte, NICOD Jean-Marc, VARNIER Christophe
From Prognostics and Health Systems Management to Predictive Maintenance 2: Knowledge, Traceability and Decision (Reliability of Multiphysical Systems Set – Volume 7)

EL HAMI Abdelkhalak, RADI Bouchaib
Dynamics of Large Structures and Inverse Problems (Mathematical and Mechanical Engineering Set – Volume 5)
Fluid-Structure Interactions and Uncertainties: Ansys and Fluent Tools (Reliability of Multiphysical Systems Set – Volume 6)

KHARMANDA Ghias, EL HAMI Abdelkhalak
Biomechanics: Optimization, Uncertainties and Reliability (Reliability of Multiphysical Systems Set – Volume 5)

LEDOUX Michel, EL HAMI Abdelkhalak
Compressible Flow Propulsion and Digital Approaches in Fluid Mechanics
(Mathematical and Mechanical Engineering Set – Volume 4)
Fluid Mechanics: Analytical Methods
(Mathematical and Mechanical Engineering Set – Volume 3)

MORI Yvon
Mechanical Vibrations: Applications to Equipment

2016

BOREL Michel, VÉNIZÉLOS Georges
Movement Equations 1: Location, Kinematics and Kinetics
(Non-deformable Solid Mechanics Set – Volume 1)

BOYARD Nicolas
Heat Transfer in Polymer Composite Materials

CARDON Alain, ITMI Mhamed
New Autonomous Systems
(Reliability of Multiphysical Systems Set – Volume 1)

DAHOO Pierre Richard, POUGNET Philippe, EL HAMI Abdelkhalak
Nanometer-scale Defect Detection Using Polarized Light
(Reliability of Multiphysical Systems Set – Volume 2)

DE SAXCÉ Géry, VALLÉE Claude
Galilean Mechanics and Thermodynamics of Continua

DORMIEUX Luc, KONDO Djimédo
Micromechanics of Fracture and Damage
(Micromechanics Set – Volume 1)

EL HAMI Abdelkhalak, RADI Bouchaib
Stochastic Dynamics of Structures
(Mathematical and Mechanical Engineering Set – Volume 2)

GOURIVEAU Rafael, MEDJAHER Kamal, ZERHOUNI Noureddine
From Prognostics and Health Systems Management to Predictive Maintenance 1: Monitoring and Prognostics
(Reliability of Multiphysical Systems Set – Volume 4)

KHARMANDA Ghias, EL HAMI Abdelkhalak
Reliability in Biomechanics
(Reliability of Multiphysical Systems Set –Volume 3)

MOLIMARD Jérôme
Experimental Mechanics of Solids and Structures

RADI Bouchaib, EL HAMI Abdelkhalak
Material Forming Processes: Simulation, Drawing, Hydroforming and Additive Manufacturing
(Mathematical and Mechanical Engineering Set – Volume 1)

2015

KARLIČIĆ Danilo, MURMU Tony, ADHIKARI Sondipon, MCCARTHY Michael
Non-local Structural Mechanics

SAB Karam, LEBÉE Arthur
Homogenization of Heterogeneous Thin and Thick Plates

2014

ATANACKOVIC M. Teodor, PILIPOVIC Stevan, STANKOVIC Bogoljub, ZORICA Dusan
Fractional Calculus with Applications in Mechanics: Vibrations and Diffusion Processes
Fractional Calculus with Applications in Mechanics: Wave Propagation, Impact and Variational Principles

CIBLAC Thierry, MOREL Jean-Claude
Sustainable Masonry: Stability and Behavior of Structures

ILANKO Sinniah, MONTERRUBIO Luis E., MOCHIDA Yusuke
The Rayleigh–Ritz Method for Structural Analysis

LALANNE Christian
Mechanical Vibration and Shock Analysis – 5-volume series – 3rd edition
Sinusoidal Vibration – Volume 1
Mechanical Shock – Volume 2
Random Vibration – Volume 3
Fatigue Damage – Volume 4
Specification Development – Volume 5

LEMAIRE Maurice
Uncertainty and Mechanics

2013

ADHIKARI Sondipon
Structural Dynamic Analysis with Generalized Damping Models: Analysis

ADHIKARI Sondipon
Structural Dynamic Analysis with Generalized Damping Models: Identification

BAILLY Patrice
Materials and Structures under Shock and Impact

BASTIEN Jérôme, BERNARDIN Frédéric, LAMARQUE Claude-Henri
Non-smooth Deterministic or Stochastic Discrete Dynamical Systems: Applications to Models with Friction or Impact

EL HAMI Abdelkhalak, RADI Bouchaib
Uncertainty and Optimization in Structural Mechanics

KIRILLOV Oleg N., PELINOVSKY Dmitry E.
Nonlinear Physical Systems: Spectral Analysis, Stability and Bifurcations

LUONGO Angelo, ZULLI Daniele
Mathematical Models of Beams and Cables

SALENÇON Jean
Yield Design

2012

DAVIM J. Paulo
Mechanical Engineering Education

DUPEUX Michel, BRACCINI Muriel
Mechanics of Solid Interfaces

ELISHAKOFF Isaac *et al.*
Carbon Nanotubes and Nanosensors: Vibration, Buckling and Ballistic Impact

GRÉDIAC Michel, HILD François
Full-Field Measurements and Identification in Solid Mechanics

GROUS Ammar
Fracture Mechanics – 3-volume series
Analysis of Reliability and Quality Control – Volume 1
Applied Reliability – Volume 2
Applied Quality Control – Volume 3

RECHO Naman
Fracture Mechanics and Crack Growth

2011

KRYSINSKI Tomasz, MALBURET François
Mechanical Instability

SOUSTELLE Michel
An Introduction to Chemical Kinetics

2010

BREITKOPF Piotr, FILOMENO COELHO Rajan
Multidisciplinary Design Optimization in Computational Mechanics

DAVIM J. Paulo
Biotribolgy

PAULTRE Patrick
Dynamics of Structures

SOUSTELLE Michel
Handbook of Heterogenous Kinetics

2009

BERLIOZ Alain, TROMPETTE Philippe
Solid Mechanics using the Finite Element Method

LEMAIRE Maurice
Structural Reliability

2007

GIRARD Alain, ROY Nicolas
Structural Dynamics in Industry

GUINEBRETIÈRE René
X-ray Diffraction by Polycrystalline Materials

KRYSINSKI Tomasz, MALBURET François
Mechanical Vibrations

KUNDU Tribikram
Advanced Ultrasonic Methods for Material and Structure Inspection

SIH George C. *et al.*
Particle and Continuum Aspects of Mesomechanics

Printed and bound by CPI Group (UK) Ltd, Croydon, CR0 4YY
16/09/2026
14963192-0001